Audrey D. Levine

 # The Sustainable Well Series
Series Editor Roy Cullimore

The Application of Heat and Chemicals in the Control of Biofouling Events in Wells, George Alford and Roy Cullimore

Water Well Rehabilitation: A Practical Guide to Understanding Well Problems and Solutions, Neil Mansuy

The APPLICATION of HEAT and CHEMICALS in the CONTROL of BIOFOULING EVENTS in WELLS

By
George Alford
Roy Cullimore

Boca Raton London New York Washington, D.C.

Library of Congress Cataloging-in-Publication Data

Alford, George.
 The application of heat and chemicals in the control of biofouling events in wells / by George Alford and Roy Cullimore.
  p. cm. -- (Sustainable well water)
 Includes bibliographical references and index.
 ISBN 1-56670-385-9 (alk. paper)
 1. Water--Purification. 2. Wells--Fouling. 3. Fouling organisms--Control--Technological innovations. I. Cullimore, Roy.
II. Title. III. Series.
TD541.A44 1998
628.1′14—dc21   98-31049
                                                                                          CIP

This book contains information obtained from authentic and highly regarded sources. Reprinted material is quoted with permission, and sources are indicated. A wide variety of references are listed. Reasonable efforts have been made to publish reliable data and information, but the author and the publisher cannot assume responsibility for the validity of all materials or for the consequences of their use.

Neither this book nor any part may be reproduced or transmitted in any form or by any means, electronic or mechanical, including photocopying, microfilming, and recording, or by any information storage or retrieval system, without prior permission in writing from the publisher.

The consent of CRC Press LLC does not extend to copying for general distribution, for promotion, for creating new works, or for resale. Specific permission must be obtained from CRC Press LLC for such copying.

Direct all inquiries to CRC Press LLC, 2000 Corporate Blvd., N.W., Boca Raton, Florida 33431.

Trademark Notice: Product or corporate names may be trademarks or registered trademarks, and are used only for identification and explanation, without intent to infringe.

© 1999 by CRC Press LLC
Lewis Publishers is an imprint of CRC Press LLC

No claim to original U.S. Government works
International Standard Book Number 1-56670-385-9
Library of Congress Card Number 98-31049
Printed in the United States of America 1 2 3 4 5 6 7 8 9 0
Printed on acid-free paper

PREFACE

Water wells form a major source of water for many communities; yet many aspects of water well management have remained unrecognized until recently. The old management practice was set in delivering specific production quotas and replacing any wells that failed to deliver an acceptable quota. Over the last twenty-five years, it has become increasingly evident that water wells actually are conduits to the subsurface universe and the organisms "down there" are capable of influencing the production characteristics of wells.

These influences are often very subtle and take time to appear and be recognized. Unfortunately, because of the nature of the organisms growing in and around wells, their activities have not been acknowledged as major impacts on the functioning of water wells. These activities extend to improving the quality of the produced water (as a result of the biofiltration occurring in the plugging zones) to degeneration in both the quality (due to sloughing of the plugs) and production (due to the plugging of the formations around the well). Essentially, the traditional approaches have been "out of sight and out of mind". The application of computer modeling based upon simple hydraulic flows using Darcy's equation, permeability units and velocity concepts is flawed by its application to simple flow systems. In water wells, ground water, and the encompassing aquifer systems, research to-date has shown the universal presence of microorganisms. These organisms, whether surviving or aggressive, will occupy surfaces and voids to impact hydraulic conductivity. This impact is mostly seen when a water well plugs as a result of microbial growth.

The subsurface biosphere clearly influences the hydraulic characteristics and has to be recognized in the practice of hydrogeology and attempts to model ground water movement. In the practice of computer modeling, errors abound as a result of the failure to recognize the true nature of the porous and fractured structures as being habitats for the growth of a diverse microbial flora. Today, it is common to build in randomizing functions such as the "Monte Carlo" scenario and it is, as it seems, a form of gambling based on ignorance rather than knowledge. It has to be remembered that a computer predictive model is only as good as the elements that are built in and as bad as the elements that are left out. The subsurface biota have been ignored to this time and this book describes some of the impacts that microorganisms have on water wells and methods of rehabilitation.

Like all new advances in science, the research into the plugging of wells calls into question many preconceived notions and simplistic solutions. For example, if there is a bacterial infestation down the water well causing biofouling, giving a one-shot heavy dose of a chlorine-based disinfectant is not the permanent solution to the problem. The work described in this book deals with some aspects of treating plugging water wells by the blended use of heat and chemical strategies. What has become very obvious during the last twenty-five years is that "no one size fits all"; "nothing lasts for ever"; and, "preventative maintenance is essential after a treatment no matter how successful the treatment has been". This means that each well should be treated as a fresh "patient", treatment should just be a part of a management strategy, and monitoring should be mandatory along with good record keeping.

This book is split into four chapters. The first chapter essentially deals with understanding the "cause" of plugging in water wells. As a result, this chapter is very theoretical (practical-minded readers may prefer to just skip it and get to the "meat and potatoes" in Chapter two). In this work, the desire of the authors is to demonstrate the complexity of the microbial world and the ways in which these microbes can plug ("throttle") water wells slowly until they can no longer deliver water in acceptable amounts (because they are "strangled").

In the second chapter, the research conducted from 1989 to 1993 using primarily the blended chemical heat treatment process (BCHT™) is described site by site. Both the successes and the failures are included. It has to be remembered that a failure often teaches more than a success! One lesson that has been learned is that each well is a new experience and each well is unique. Also, there is always a need to be adaptable to the differences that can occur between any two wells, whether they are ten feet or ten miles apart.

Chapter three is a transcription of a one-day workshop given by George Alford in February 1998 in Regina, Canada. This chapter provides an up-to-date anecdotal evaluation of the methodologies for treating plugged water wells, with emphasis on the BCHT™ process. While this process is the focus of the chapter, this does not mean that there are no other effective methods for treating plugged wells. Perhaps the most fundamental lesson that can be learned here is that the application of heat in the treatment of water wells, regardless of the chemical strategy, can offer advantages when performed in a realistic manner.

In the final chapter, there are three parts. The first is essentially a question and answer period in which George Alford responded to questions and comments from the floor of a workshop. This workshop was organized as a part of a series on sustainable water wells by the Regina Water Research Institute of the University of Regina. The second part of the chapter covers final comments by the authors, partly in response to the questions from the floor and partly as a general summation of the state of the art. In the third part of this chapter, there is a description of a parallel subsurface extreme oceanic environment where plugging microorganisms also thrive. This was at the site of the *RMS Titanic* that is now covered with rusticles that are extracting iron from the steel hull. The dive was made by Roy Cullimore on August 16, 1996 and included the recovery, growth and analysis of rusticles from the ship. Parallels between the plugs down water wells and the rusticles are discussed.

The authors would like to acknowledge the following for their support in the development of the studies, and the production of this book: Roy Leach and Steve White (U.S. Army's Corps of Engineers), Bill Rogers (co-inventor of the BCHT™ process), John Lebedin (PFRA-TS, Canada Agriculture), Brent Keevill, Twyla Legault, Lori Johnston, Wade McLean (Droycon Bioconcepts Inc.), and Natalie Ostryzniuk (Regina Water Research Institute) for the diligent transcription of ideas to publishable text. Additionally, the authors would like to acknowledge the financial support of the U.S. Army's Corps of Engineers, Canada Agriculture (PFRA-TS), the National Research Council of Canada (IRAP program), the Natural Sciences and Engineering Research Council of Canada, and the Saskatchewan Research Council.

George Alford
Roy Cullimore

October, 1998

ABOUT THE AUTHORS

George Alford

George has been involved in the ground water industry for over 30 years. This has involved a hands-on approach working with private and municipal water supply systems throughout the U.S. and Canada. His interest in research was stimulated by the severe local water well biofouling problems that were occurring. This has led to serving on two regional EPA ground water panels and becoming co-chairperson of the "First International Symposium on Biofouled Aquifers" held in Atlanta, 1986 funded by the EPA. George has been directly associated with research projects at six North American universities and holds seven patents jointly with research associates from these projects. George has been the principal investigator in seven research contracts funded by the EPA and/or the U.S. Army's Corps of Engineers and has co-authored ten technical papers. He has worked at many sites such as Shaw AFB, USACE/REMR sites 1 and 2, New Lyme Landfill, Bofers-Nobel Superfund Site, and the Rocky Mountain Arsenal. Since beginning operations, George has rehabilitated over 2,000 severely plugged wells successfully and has introduced the Alford/ARCC preventative maintenance programs which have been adopted as a standard management procedure at numerous hazardous waste sites across the U.S. George also acts as a consultant to the USGS and the Centers for Disease Control.

Roy Cullimore

Roy is an applied microbial ecologist trained at the University of Nottingham, U.K. Since 1975, he has been Director of the Regina Water Research Institute at the University of Regina. George Alford (co-inventor) and he have five patents including the biological activity reaction test (BART™) and the blended chemical heat treatment (BCHT™). Roy has published over one hundred refereed papers, two hundred and seventy technical reports, and has received over 3.5 million dollars in research funding. Now he is President of Droycon Bioconcepts Inc. of Regina, a biotechnology company involved in research, development and manufacture. He authored *Practical Manual of Groundwater Microbiology* which was published by Lewis Publishers in 1993. Currently, Roy is editor of a series of books on sustainable wells. He is presently involved in research on the rusticle growths on *RMS Titanic* and dove to the ship in 1996 and 1998 as a part of the Discovery Channel expeditions. Also, Roy is currently involved in the AWWARF water well rehabilitation project being undertaken by Leggette, Brashears and Graham, Connecticut.

CONTENTS

1. THEORETICAL PERCEPTIONS OF A FUNCTIONING WATER WELL 1

I. INTRODUCTION 1

II. EARLY DIAGNOSTIC TECHNIQUES FOR FAILING WELLS 2
    A. Symptoms 3
        1. Losses in Flow 3
        2. Increases in Drawdown 3
        3. Generation of Cloudiness 4
        4. Color in Water 4
        5. Taste and Odor Problems 5
    B. Traditional Rehabilitation Techniques 8
    C. Recognition of Biofouling - Intrinsic Concern 10
        1. Water Production 11
        2. Chemical and Physical Characteristics 11
            a. Iron (Fe) 11
            b. Manganese (Mn) 12
            c. Total Suspended Solids (TSS) 12
            d. pH (acidity and alkalinity) 13
            e. Redox (reduction-oxidation) 14
            f. Temperature 15
            g. Freezing ($<0°C$) 16
            h. Total Dissolved Solids (TDS) 17
            i. Nutrients 17
        3. Biological Challenge Determination 19
    D. Recent Technological Advances 20

III. THEORETICAL APPRAISAL: CAUSES OF BIOFOULING 22
    A. Major Symptoms of Biofouling 23
    B. Plug Formation 24
    C. Sequence of Symptoms in Plug Formation 27
    D. Recalcitrant Chemical Accumulates 28
    E. Compromised Water Quality 30
        1. Iron 30
        2. Total Organic Carbon 31
        3. Total Nitrogen 31
        4. Total Phosphorus 32
        5. Redox Potential 33
        6. Temperature 34
        7. Microbiological 34
    F. Production Failures 34
    G. Appraisal of the Causes of Biofouling 34

IV. CHRONOLOGICAL SEQUENCES 36
  A. Water Production 36
  B. Biofouling Process 37
    1. Initial Colonization 37
    2. Primary Void Volume Occupancy 38
    3. Primary Stabilization 38
    4. Secondary Void Volume Occupancy 38
    5. Plugging 38
    6. Total Plugging 39
  C. Critical Indicators: The Stages of Biofouling 39
  D. Chemical Indicators Reflecting Biofouling 40
  E. Physical Factors Influenced by Biofouling 41
    1. TSS 42
    2. Mean Particle Size (microns) 43
    3. Percentage Volume Distribution to the Allocated
       Sizes 44
    4. Surface Area and Wet Weight 44
    5. Temperature 44
    6. Pump Rates and Draw Down 45
    7. Redox 46
    8. Direct Visual Inspection 47
       a. Free-Floating 47
       b. Mucoid Tubercles 47
       c. Hardened Plates 48
       d. Covert Plugging 48
  F. Biological Factors Influenced by Biofouling 48
    1. Attached Habitat 49
    2. Oligatrophic Nutrient Regime 49
    3. Erratic Population Recording 50
    4. Specialized Growth Requirements 50

**2. APPLICATION OF HEAT AND CHEMICALS IN THE CONTROL
   OF BIOFOULING EVENTS IN WELLS** 51

I. APPLICATION OF HEAT, BACKGROUND INFORMATION ... 51

II. HISTORICAL ASPECTS OF THE APPLICATION OF HEAT ... 54
    1. Clogging 54
    2. Plugging 55
    3. Bioplugging 55

III. PASTEURIZATION OF WATER WELLS 57

IV. BLENDING CHEMICALS WITH HEAT TREATMENT 59

V. BCHT™ DEVELOPMENTS 62
    A. Upper Woods River, 1988 62
    B. Brookville Lake, 1990 64
    C. Shaw Air Force Base, 1992 66
    D. Garrison Dam, 1992 70
    E. Leesville, 1992 75
    F. Bofers Site, 1993 83
    G. Mississippi River, 1993 85
    H. Conclusions, 1988 to 1993 94

3. REVISIONS AND RETROFITTING, BRINGING DOWN THE COST 97

I. INTRODUCTION 97
    A. Application of Jetting 103
    B. Selection of a Suitable Heat Source 106
    C. Problems: Heat and Chemical Applications 107

II. SUCCESSFUL REHABILITATION OF RELIEF WELLS 109

III. RISK ASSESSMENTS AND ENVIRONMENTAL IMPACT ... 110

IV. BCHT™ TREATMENT 113
    A. Applications Protocol 113
    B. Hazardous Waste Sites 116
    C. Hot Versus Cold Treatment 118

V. DIAGNOSIS OF BIOFOULING WELLS 119

VI. PREVENTATIVE MAINTENANCE 123

VII. TARGETING THE BIOFOULING 125

VIII. HAZARDOUS WASTE SITES 126

IX. WELL BIOFOULING 128
    A. Historical Background 128
    B. The Coliform Problem 129
    C. Good Record Keeping and Monitoring Performance 131
    D. Preventive Maintenance, The Essential Component 132

4. DISCUSSIONS 135

I. QUESTIONS AND ANSWERS 135

II. SUMMARY	151
III. FINAL COMMENT	157
IV. *TITANIC*, The Connection Between Rusticles and Clogging	162
APPENDIX, Definition of Terms	167
SELECTED BIBLIOGRAPHY	175
INDEX	177

1

THEORETICAL PERCEPTIONS OF A FUNCTIONING WATER WELL

I. INTRODUCTION

Water wells have been traditionally viewed essentially as a pipe that passes down into the ground and fills, at least partly, with water. This water can be drawn from the well by some form of pumping action and is replenished by ground water flowing towards the well. Definitions of well include "a shaft sunk into the ground to obtain water... a mineral spring... a spa... a water-spring or fountain". Water coming from a well is generally considered by many people, since it has come from under the ground (subsurface), to be free from biological activity. Problems arising from water wells are, therefore, often considered to be physical and/or chemical problems and are treated as such. In reality, the subsurface environment is rich in biological activity that can affect the performance of water wells in many ways. This document will address some of these mechanisms which range from the good and the bad to the ugly.

The good activities reflected by a well is its ongoing and reliable production of water of an acceptable quality and quantity when demanded. This demand may be created by pumping water from the well (production well), water under a head pressure springing from the well (relief well), or, through pumping water back down a well (recharge or injection well). It is generally understood that a water well will continue to perform satisfactorily if its performance matches the level obtained when the well was first brought on-line (developed).

Failure in water wells is usually related to either: the physical failure of the structure (stress-, corrosion-, or plugging-related), loss in ground water recharge to the well, or deterioration in the product water quality (chemical) and quantity (physical). These failures have been viewed traditionally as not involving a biological factor. Today, there is a growing body of evidence that supports the fact that the subsurface is, in fact, richly endowed with a wide variety of microorganisms. Such microbes can influence the efficiency of water wells when they are present within the zone of influence of the water well. Very often operators view the obvious occurrence of bacteria (e.g., the presence of coliform bacteria in hygienically compromised wells) as having arisen by contamination from the surface (e.g., through leaking from septic wastewater systems). Still today there are many operators of water well systems who believe that their well systems are essentially "sterile" and do not, therefore, need to be concerned about any possible biological challenges to the integrity of their system(s). In reality, microorganisms are ubiquitous (i.e., virtually everywhere), and it is a matter of whether the local environment is suitable for them to become active. A water well often provides a suitable environmental "setting" for microbial activity because of the changes that occur in flow rates, oxygen concentrations, and the types of surfaces that are present.

There are three exercises which are related to developing a long-term management of biofouling: 1. recognizing the potential microbial challenges to the wells; 2. rehabilitating biofouled (or biologically challenged) water wells; and 3. establishing a preventative maintenance (PM) program to ensure sustainability. Such initiatives should allow a prolonged life to the water well installation and, therefore, very significant economic advantages to the user.

Reacting to a Problem with a Water Well:

- ❑ Observe the symptoms
- ❑ **Determine the cause (microbial, chemical, or physical)**
- ❑ Develop a method to remediate the problem
- ❑ Apply the remedial practice
- ❑ **Determine whether that treatment has been successful**

- ❑ Develop and apply a monitoring program
- ❑ React quickly to control any repetition of the problem

Figure One

II. EARLY DIAGNOSTIC TECHNIQUES FOR FAILING WELLS

Failure represents the water well is no longer delivering an accept-able water quality in adequate amounts to meet the product goals expected of that well. Early diagnoses were based on the user observing deterioration in production that could sometimes be dramatic. Typical events include: losses in flow; serious increases in drawdown to achieve flow; generation of cloudiness (turbidity) and/or color in the water; appearance of taste and/or odor problems; and increasing hygiene problems. Early diagnosis has commonly been based upon simple techniques that relate to the problem(s) the user had observed. These techniques are listed below.

A. SYMPTOMS
1. Losses in Flow

There is a gradual diminishment in the amount of water flowing from the well over a given time. This is usually recognized by a lengthening in pump times to achieve a given production goal, or reduced flow rates from the water well system. The user may observe this by the ongoing casual monitoring of the well. In general, the cause of the failure may be put down to two possible events. First, the water level declines in the ground water so that it can no longer service the well efficiently. Second, a plugging process has prevented water from reaching the borehole in sufficient volumes to service the demand.

2. Increases in Drawdown

One symptom often observed by the user was the fact that the water level in the well was stabilizing at deeper and deeper positions during pumping. This meant that the pump was only able to draw adequate water by the creation of a larger water head drawing water into the borehole through a cascading effect. The loss in water at the start of pumping is referred to as the "drawdown"; and it is generally recognized that, where this occurs, there is a plugging process that is impeding flow to the well. If the user notices that the (static) water level when there is no pumping is falling, this may be taken to indicate that the water table is falling in the aquifer. In this case the user may suspect that the ground water system is being overtaxed (e.g., output is exceeding input). Where plugging is suspected, it has traditionally been thought that this was largely as a result of silting up (physical phenomenon) and/or encrustations (chemical phenomenon) forming to impede the entry of water into the well. Silting up has generally been thought to be caused by the entry of silt particles into the well that would reduce water entry and also, settle in the borehole to reduce the capacity of the well.

3. Generation of Cloudiness

Cloudiness in water was generally thought to represent degeneration in the chemistry of the water that caused various salts and colloids to become suspended in the water. The cause of such cloudiness was considered to have been a result of the well water reacting between different strata of water. This was also often associated with the formation of encrustations and/or plugging within the well. The usual test was to hold the water up to a light and observe the clarity. It was not realized initially that water clarity did not necessarily correlate in any respect to any microbial presence. It is now known that a completely clear water sample can still easily contain 100,000 organisms per ml, and that cloudy waters are very likely to contain significant populations of microorganisms.

4. Color in the Water

Water may become colored if there are artifacts present which are colored. These artifacts may range from the molecular to the colloidal and the particulate. Color, when it appears in the product water, is indicative of a problem becoming established which may challenge the ongoing operation of the well. Colors sometimes observed in water range through the following range: yellow, orange, brown, gray and black. At the same time the water may retain a high clarity or become clouded. When the water is left to stand at room temperature, the color and the cloudiness may intensify before the material settles out to the bottom of the container as metallic floaters, or as thin, loosely packed wooly deposits. One test method uses a drop of corn whiskey to speed up this process and is used as a positive indicator for the presence of iron-related bacteria.

<u>Yellow</u>, <u>orange</u> or <u>brown</u> waters are commonly associated with the presence of iron in the water at concentrations of greater than 0.5 parts per million (ppm). Yellow waters tend to retain a high clarity while orange waters may generate some cloudiness. Brown waters are generally clouded and may contain a significant load of colloidal and particulate iron. This event has traditionally been linked to the onset of serious iron-plugging problems.

<u>Gray</u> water is a dirty cloudy water which often has a high microbial population. It has usually been associated with high levels of organic challenges of a type that might occur if a septic wastewater or an organic contaminant plume were to enter the zone of influence of the well.

Black water is a relatively common occurrence particularly when wells are activated after a long period of inactivity. Usually the black color is generated by the production of black iron sulfides. This commonly occurs in waters that have become infested with sulfate-reducing bacteria (SRB). There is little or no oxygen in the water. Sometimes these black deposits form into granulated or thread-like masses that float within the water.

Cloudiness and Color in Water are Signals that:
- **Microbial growths may be occurring in the water**
- **Any slimes and encrustations in the wells are sloughing**
- **Production problems (plugging) are likely to increase**
- **Taste and odor problems are more likely to occur particularly if the well's production schedule is radically changed.**

Figure Two

5. Taste and Odor Problems

Most water problems are first perceived by their odor rather than by the taste generated. Odors range widely and are sometimes difficult to categorize in a manner that other people would easily understand. Common problems have been reported associated with the following odors: rotten eggs, fishy, septic, kerosene-like, earthy-musty, vegetable, fruity and skunky. Each of these odors can be related to some biological event associated with the well.

Rotten egg odor is perhaps one of the most common and relates to the decomposition of organic matter and the reduction of sulfates (by SRB) to hydrogen sulfide. This hydrogen sulfide has a strong odor that is also present in the "black rot" of eggs. Hence, the common name for this odor in water is the rotten egg odor. It occurs commonly when a well has been shut down for a prolonged period and there is little oxygen in the ground water and/or a lot of organic material.

Fishy odors are generated by a group of bacteria called pseudomonads (soo'do-mo'nads). They grow where there is oxygen and often form slime-like growths (biofilms). It is slimes dominated by these bacteria that commonly grow over the scales of fish and give the fish that same characteristic smell. When these odors are observed, the well waters are probably oxygenated and also, carry a range of organics that these bacteria can use for growth.

Septic odors relate to the presence of sewage and septic tank wastes but can also be generated during the anaerobic (oxygen-free) degradation of organic material. Generally, the water may also show a moderate amount of cloudiness and sometimes may have a gray color. Fresh sewage tends to have a distinctive fecal odor (manure-like) but septic wastes are more difficult to describe since the odors are more transient (vague) and less definable. Normally, the presence of septic odors should trigger a testing of the water for the presence of coliform bacteria. This is because the coliforms have been used as the indicator organisms for the presence of fecal material, and hence indicate that there is a hygiene risk when they are detected (one coliform organism per 100 ml is considered by many regulators as the borderline maximum acceptable number in water).

Kerosene-like odors can occur and may be due to the presence of hydrocarbons (e.g., gasoline, jet fuel) in the water, but there are also other bacteria which can generate similar odors. The most commonly recognized group that can do this is the pseudomonad bacteria.

Earthy-musty odors are very similar to those emanating from freshly turned soils. The reason for this similarity is that one group which generates this odor occurs widely in soils and often occurs growing in the unsaturated regions associated with ground water are the streptomycete bacteria. These grow as threadlike growths over surfaces and are very competitive, often dominating the other microbes at the sites. This is because the *Streptomycetes* also are a source for a range of antibiotics. Streptomycin is one of these antibiotics. They are aerobic and require oxygen for growth and also degrade a range of organics. As a result of this, these streptomycete bacteria may be found in recharge zones where there is a high organic potential and the water remains oxygenated. Here the earthy-musty (geosmin) odors may be created and be carried in the water. Also carried in the water will be a large number of spores that may be detected when the water is examined for microbiological content.

Vegetable odors are generated by some bacteria but are more commonly associated with the growth of micro-algae and blue-green algae (now known as the cyanobacteria). These organisms usually require light to grow (for photosynthesis); and so it may not be commonplace to obtain vegetable-like odors from well water unless there has been a very shallow local recharge; or the borehole water column is supporting the growth of these plant-like microorganisms. It should be remembered that some of these organisms are able to compete in the dark with other non-photosynthetic microorganisms and could generate these odors.

Fruity odors are very uncommon in wells and are usually associated with the presence of yeast capable of producing a range of fruity esters during growth. Yeasts are commonly not reported as a normal part of the ground water flora; their presence would indicate that a high carbohydrate or organic acid pollution might be occurring. Some bacteria do produce similar odors but it has been found through practical experience that the product ground water does not possess the odor, but that it appears when the bacteria are cultured.

Skunky is a description used by many simply for an unpleasant odor to which the operator cannot ascribe a particular smell. Generally, the odors are septic or earthy-musty in nature.

Odors, when they are reported, usually are associated with some malfunctioning in the system. The general attitude is, first, that the water has become polluted with chemicals that bear this smell. It is not often recognized that these odors may also be a signal that microbially induced fouling is increasing. A good example of this is the generation of fish-like odors (during the early growth of pseudomonad bacteria in aerobic conditions) and septic odors later as the oxygen is depleted and the bacterial flora changes to an anaerobic type. The presence of odors in the ground water can be taken to indicate that a problem is arising which may be microbiological in nature.

Water Wells are Changing

From being:
- Disposable
- Cheap to replace
- Minimal in maintenance requirements

To being:
- Expensive to replace
- Sustainable
- Requiring ongoing preventative maintenance

Figure Three

B. TRADITIONAL REHABILITATION TECHNIQUES

Historically, there have been three major reactions to the declining operation of a water well of any common type. One reaction (which may be popular with some well drillers) is to simply abandoned the well and install a new well of similar or greater capacity to compensate for the abandoned well. A second reaction involves attempting to change the operating techniques (e.g., pump times, volumes, sequences of up and down times, and control flow by drawdown limitation) or change components in the well (e.g., pump, screen) in the hope that the well will recover from the observed problem. The third reaction is to attempt an analytical approach to the problem. This involves determining: first, the cause; second, confirm that the effects witnesses can be related to the cause identified; and three, determine and apply a treatment strategy that will counteract the cause allowing the well to function in its designed manner.

The increasing economic and environmental costs and concerns are now restricting the ability of a well user to simply replace a failing well. Economic concerns are relatable to the increasing costs involved in well replacement. There is also a growing sensitivity for maximizing the use of each well installation through extending its useful life. Environmental concerns are being brought to the fore by the fact that ground waters are no longer being seen as an infinite resource that can be exploited without limitation. In some areas, the aquifers are now being heavily depleted by the present demand and there is little ability to provide additional capacity. Another major environmental concern is the impact of various forms of pollution on the well fields.

In the decades gone by, the general attitude may be summarized by the "out of sight, out of mind" mentality. It was commonly thought that ground water should be given a lower status of concern to surface waters. Various chemical leakages from industry, agriculture and the various service industries were not considered to be of as much importance as in surface waters. When a pollutant impacted on a surface water, the effects could often be relatively quickly appreciated through such effects as radical eutrophication, deteriorating water quality and the water becoming unacceptable to the user. One major difference between surface waters and ground waters is the fact that the former flows as large unconfined masses, while the former moves as a confined mass within various porous media. This difference is very critical to the current understanding of ground water flows and quality.

It is not easy to appreciate the complex interactions which occur between the flowing ground water and the media it is passing through as it moves to a well, a spring or interfacing with another aquifer. It has been generally believed that ground waters are essentially sterile (devoid of biological activity) and that activities within an aquifer may be explained almost exclusively by a combination of physical and chemical processes.

Today, hydrology in ground water systems leans heavily upon this assumption. As subsurface microbiology (the study of microorganisms in the crust of the planet) it is becoming increasingly evident that ground water movement and quality is affected by microbiological interactions. In the past decades these have been ignored and one of the major consequences of this has been that the role of microorganisms as biological filters (interface) has been ignored.
Pollutants within a ground water system may become entrapped (and possibly degraded) within these "biological filters" and so not appear in the ground water resurfacing through a well.

Environmental monitoring of the water product from a well may not necessarily give an "accurate" picture of the chemical loading in the transient water itself. Essentially, there has been a tendency for ground water users to rely on the product (bio-filtered) water for environmental assessment, and yet this water may not accurately allow a risk-assessment for that well (due to the bio-entrapment of some chemicals of concern). In the next two decades, this realization may cause much tighter environmental constraints on new well installations. Therefore, a greater attention would be paid to extending the service life of existing wells through preventative maintenance and effective rehabilitation programs.

The mindset focussed on the fact that water wells are physical objects set within the chemical and physical world has generated traditional attitudes. One effect of this is the consideration that the dysfunctional well is a result of chemically/physically driven corrosion, encrustation, plugging or the physical collapse of the system through such events as "silting up" and "collapsed" aquifer and well structures. Acidization was commonly applied to dissolve and disperse the plugs and encrustations while various disinfectants (such as different formulations of chlorine) were used to control any coliform and other bacteria that may be growing down the borehole. Slime formations were considered by many to be simply physical-chemical accumulates which may result in plugging, encrustation and corrosion occurring.
Even today, the camera logging of a water well is thought to be sufficient to view all of the biological and much of the chemical

deposits (e.g., silts and salts) which can be causing problems around a well. Combinations of disinfectants, selected acids and even, in more recent times, dispersants have become a part of the arsenal of weapons being used to rehabilitate problems in a well.

One of the findings from these reaction scenarios has been that "no one size fits all" and that each well should be treated as unique, requiring customization of the parameters to optimize the maintenance practices. This approach stems from observations that each well can be characterized as being different from the other wells in the same field. Indeed, there are many experimental experiences where two wells of the same construction and characterization placed within feet (meters) of each other, in supposedly the same aquifer formation, bear very different characteristics. An unfortunate result of this is that a treatment scenario may be successfully applied to one well in a field, but that same treatment may fail on a neighboring well with exactly the same characteristics in its construction, operation and mode of failure.

Each Well should be treated as a Separate Installation because:
- **No one size fits all**
- **Each well may be set in different environments even though the surface terrain may look very similar**
- **Wells close together do not have to have identical characteristics**
- **Wells in a well field will be influenced in unique ways by the activity of other wells in that field**

Figure Four

C. RECOGNITION OF BIOFOULING - INTRINSIC CONCERN

Today, more and more wells are being subjected to an analytical determination of the nature of the problem causing concern. This analytic approach has now gained a level of acceptance operating at three levels:

1. Water Production
2. Chemical and Physical Characterization
3. Biological Challenge Determination.

1. Water Production

Water production is usually determined primarily by the volume of flow from the well achieved over a given time period (e.g., hour, day, etc.). Where a well is operating routinely below its capacity to produce, the volume of water flowing from a well may continue to appear optimal (as demanded) while transmissivity to the well may be becoming impaired.

For example, a well designed to pump water from the well at 50 gpm may actually be capable of flowing at 500 gpm. It is, therefore, operating at only 10% of its theoretical limit. If that well becomes plugged, for whatever reason, then the flow into the well will become reduced. Plugging would have to cause a 90% decline in flow before the pumped volume from the well would be affected. In the field, it has been observed that wells may suddenly lose production when the capacity is down to 20 to 60% of the original theoretical capacity. In the well given as an example, the flow from the well may decline dramatically simply because the plugging would have been nearly complete before the well flow was affected. Early signals for loss in production include increases in drawdown during pumping and longer times for the recovery of the water column to its standard (water table related) position after pumping. Pump tests can stress the well to its maximal production capacity. This forms one way in which the production capacity of the well can be examined. If these pump tests are performed routinely, the rate of production capacity loss can be projected and a preventative maintenance program initiated.

2. Chemical and Physical Characteristics

Very often when a water well is becoming compromised by some adverse process, the quality of the water changes in terms of its chemical and physical characteristics. These shifts have been used as signals that a well may be becoming impaired. Some of the common characteristics that have been used are listed below.

a. Iron (Fe)

Iron is accumulated in plugging and other forms of biological interfaces. As these accumulates grow, the water quality may reveal only traces of iron since much of the iron is being trapped before the water enters the borehole and is pumped from the well. However, as the biological mass continues to grow, it becomes periodically unstable. During these periods the mass begins to break up and slough into the pumped water. This becomes reflected in the higher iron content in the water. Erratic changes in the iron concentration of water being pumped from a well, therefore, can be used to indicate that the form of bioaccumulation (e.g., plug) is now maturing to the point of becoming unstable. This may be readily observed in the water through the generation of colors (yellow, red to brown), turbidity, and copious red to brown slime deposits forming in static water tanks and along slow flow pipes. Increases in problems with water treatment equipment such as filters and softeners may be noted at this time. Generally, the critical range of concern for iron is between 0.1 ppm (no obvious problem) and 1.5 ppm (obvious problem).

b. Manganese (Mn)

This is another chemical of concern that is often found in unacceptable levels in product waters from a well. Manganese, like iron, can also sometimes bioaccumulate in the plugs and other biological interfaces around a water well. While iron tends to collect relatively close to the well, manganese tends to be more distributed into the formation. Most wells show much lower levels of manganese than iron and the ratios (Fe: Mn) tend to remain fairly stable. Some typical ratios that are seen in wells range within approximately 100:1, 30:1, 3:1, 1:1, 1:10. Where there is more manganese (e.g., lower only in last ratio), the color of the water tends to be different ranging through shades of gray to black. In general, waters with higher manganese contents (relative to iron) are more difficult to apply to a preventative maintenance program. The critical concentration ranges of concern for manganese are between 0.01 ppm (not normally a major problem) and 0.5 ppm (obvious problem).

c. Total Suspended Solids (TSS)

When a well has been biofouled to the point that sloughing is occurring, the suspended solids content in the pump becomes more evident. It may be seen as a reduction in clarity, increase in turbidity, increase in cloudiness, presence of visible particles, and/or the development of deposits (loose, granular or slimy) when the water is left standing. These suspended solids may range from silt and clay particles to colloidal particles. To stay in suspension, these solids have to be very small in size (usually they range from 50 or 60 microns down to less than a micron). A micron is one thousandth of a millimeter or around a quarter of a millionth of an inch. Bacteria range in sizes of around a millimeter or two. Colloids are "jelly-like" particles that are composed mainly of water but bound together by bacterial or chemical polymers. These are long string-like interwoven molecules that have the ability to attract and hold water. Very often bacteria and other microorganisms are suspended and "travel" within these biocolloids. When slimes begin to slough for whatever reason, biocolloids are released and move through the water. Laser particle sizing is becoming an excellent tool for estimating the volume, size and even the shape of these particles. Some problems have been experienced because the biocolloids are very mobile and appear and disappear from within the laser-scanning beam in a matter of seconds.

d. pH (acidity and alkalinity)

pH is an easy parameter to measure and while it cannot detect biofouling, it can indicate whether conditions are likely to support biological activity. In reality there are some microorganisms that can flourish at any pH in the range of 0 to 14, but as one approaches the very acid (less than 4.5) and very alkaline (greater than 11.5) then only a few very specialized organisms can flourish. The pH that supports the greatest diversity of microbial growth lies in the neutral to slightly alkaline range from 7.2 to 8.8. Some microorganisms are actually able to "buffer" the pH to optimize it for their growth. Often this "ideal" pH is at 8.3 to 8.7. Taking a pH reading can, therefore, give some information about the diversity of the microflora that may be biofouling the water. As a rule of thumb, the following can be used to determine the possible influence of pH:

> <u>Less than 4.5</u> microflora may be quite restricted. If there are oxygen and sulfates present, there may be a potential for microbially generated acidic leaching to be occurring (*Thiobacillus* is a common species causing this). Remember that many slimes are capable of continuing to grow and cause biofouling even in these acidic regimes. This is because the surface coating of the slime can buffer the pH up to a more accommodating range for the microorganisms. In the laboratory, biofouled porous media columns have been observed to return to a neutral pH after an acidic shock (e.g., at pH 3.5) due to this buffering action. pH elevations of 0.2 to 0.8 pH units per hour have been observed.
>
> <u>4.5 to 7.1</u> the microflora is traumatized to some extent and, generally, the process of biofouling may become retarded

(particularly below pH values of 5.5). There may be a less diverse microflora associated with the biofouling that is occurring.

7.2 to 8.8 is optimal for the growth of many microorganisms and the pH will not act to impede their activity.

8.9 to 11.4 very little is known of the effects of mildly alkaline conditions on the activity of slime forming microorganisms. Generally, the activity of the microorganisms becomes suppressed and the slimes may tend to thicken (presumably as a protective function to the rising pH).

Greater than 11.4 tends to rapidly retard microbial activity in the water phase, but little is known about the impact on the micro-organisms protected within a plug or slime.

e. *Redox (reduction-oxidation potential measured in millivolts)*

Redox is becoming a very good indicator of the potential for microbial fouling to occur. Very frequently, microbial activities concentrate around the redox front which is formed where water is moving from a reductive (negative millivolts of lower than -50) to an oxidative (positive millivolts of more than +150). The water in the bore hole may be a reflection of just where the microbial activity is concentrated but it must be remembered that the reading being obtained is usually of the product (post-diluvial) water which may already have passed through a zone of biofouling. The redox of this water can, however, prove to be a very useful guide:

Greater than +150 would indicate that the water may have passed through a redox front and that the biofouling may be concentrated back further in the formation. If the redox value is much higher than +200, then the whole ground water system around the borehole may be oxidative. In these circumstances, it is possible that a massive but dispersed type of aerobic fouling may be occurring which is very difficult to control.

-50 to +150 would suggest that the redox front is close by, or within, the borehole itself. This would be a potentially serious circumstance because the microbial fouling may be occurring within the pump, on the screens and in the gravel pack. At these sites, there is a greater possibility that the production capacity of the well could be seriously impaired by plugging or a dramatic and unacceptable reduction in water quality.

-50 to -200 represents virtually oxygen-free (anaerobic) water with a greater probability for corrosion (from SRB activities), offensive

taste and odor generation and a poor water quality. Another serious concern for these waters is that there would be a high probability that downstream from the wellhead, conditions would become oxidative. For example, at an aeration unit in a filtration plant, as the redox conditions rise to within the range of -50 to +150, a massive amount of microbial activity may be focussed at that "shift point" with the result of a major biofouling.

<u>Less than -200</u> are generated by extremely reductive anaerobic conditions in which microbial activities may tend to be dispersed. Any microbial activities within the formation and the borehole may be very limited. However, if downstream the redox rises to a more oxidative regime, microbial biofouling could become focussed at those redox fronts where they form.

f. Temperature

Since we humans function internally most happily at 35 to 37°C, there is a general belief that everything else should also grow best at that temperature. In ground waters, one major factor affecting the microorganisms is the fact that the temperature of the water remains relatively constant (unless a local recharge of some type is occurring). Like pH and redox, temperature does influence the rate and form of microbial biofouling, but, provided the water remains liquid, there is always at least microbial presence that could trigger a fouling process. The following is the potential influence of temperature on the type of microbial activity:

<u>25 to 32°C</u> is generally considered to be the range that will support the greatest diversity of microorganisms. Many of these microbes will function most efficiently at between 27 and 30°C and very often it is other factors that limit the activity level.

<u>32 to 42°C</u> is a temperature range similar to that found in warm blooded animals. Not surprisingly, therefore, some of the bacteria which are able to grow within animals are also able to thrive in ground waters at these temperatures if other conditions are suitable. Activities may be very rapid at these temperatures although the range of microbes able to function will begin to become restricted over the range of 39 to 42°C. It is interesting that the upper end cut-off temperatures for these organisms are often over a very narrow range for the microbes in this group.

43 to 75°C are a range of temperatures where only specialized (thermotrophic) microorganisms can remain active. As the temperature rises, so the range of microbes continues to narrow. The sequential forms of microbial growth that occur around hot sulfur springs are a good example of this form of restricted range of growth.

75 to 160+°C is a very restrictive range for organisms to remain active. While the boiling point of water is 100°C at sea level under normal atmospheric pressure, the hydrostatic pressures in the (deeper) ground waters is such that the water does not boil until much higher temperatures. For example, at 500 meters, the hydraulic pressure would reach 5 MPa that would cause the water to have a boiling point of 260°C. There is a growing body of knowledge that is supporting the hypothesis that microorganisms can remain active at greater than 100°C provided the water remains liquid. The massive microbial biofouling of sunken wrecks (e.g., the Titanic) and massive populations recorded in deep oceanic sediments (e.g., 10 tons per hectare in the top 100 meters of deep Pacific Ocean sediments) support this potential.

15 to 24°C is a temperature range over which many of the microbes that function well at 25 to 32°C begin to shut down. These are called the mesotrophs. Some bacteria are able to adapt to lower temperatures and so remain active (facultative psychrotrophs). There is generally a reduced level of activity in the lower part of this range of temperatures and a more restricted range of microorganisms is recovered.

5 to 15°C supports the activities of microorganisms that are able to survive and grow at low temperatures (psychrotrophs). These organisms are usually somewhat slower growing and often take a prolonged time to adapt to the temperatures. Below 8°C, the range of microorganisms able to remain active is reduced and the rate of activity becomes slower. Below 5°C there is a marked restriction in the range of psychrotrophs and biofouling events are commonly slower in their formation.

g. *Freezing (<0°C)*

Water does not freeze evenly. Water is expressed from the biocolloids and other particles in the water to form ice. The amount of (liquid) water retained in the particle becomes reduced but remains liquid. Microbes present within these particles can survive and possibly function while in these particles. If the water is saline then it does not freeze at 0°C, and so microbial biofouling activities can continue albeit very slowly.

h. Total Dissolved Solids (TDS)

We always consider that there is a real difference between fresh- (<0.1% salt) and sea- (>1.0% salt) water because of the difference in the dissolved salt concentration. Microorganisms have a very different salt tolerance than we do. In ground water, the (total) salt concentration is measured as the total dissolved solids in parts per million. The TDS ranges that affect microbial activities are very different to those that affect plants and animals growing on the surface of the planet. Some critical ranges are discussed below.

Greater than 12% TDS restricts the range of microbes to those that can tolerate the very high osmotic pressures created by the salts. These bacteria are called the halotrophic bacteria and appeared to have evolved early in the life history of the planet. Various strains can be active at various salt concentrations right up to saturated salt solutions (brines) in which crystallization is occurring. In waters that are oxidative or at the redox front, growths of these organisms will sometimes generate red slimes.

5 to 12% TDS has an impact on the variety of microorganisms that are able to survive and grow. As the salt concentration rises over this range, fewer and fewer strains of bacteria are likely to be recovered from a biofouling event. The rate of biofouling may be affected particularly over the range of 7 to 12%.

0.1 to 5% TDS supports the growth of most microorganisms and biofouling will not commonly be restricted by the TDS. There is likely to be a broader diversity of microflora over the lower part of the range (i.e., 0.1 to 1.5% TDS), but this may not necessarily impact on the biofouling event that is being generated.

10 ppm to 0.1% (1,000 ppm) TDS does begin to have an impact on the diversity of the microflora that are able to function. Generally such waters would also be very low in nutrients and not be supportive of massive microbial biofouling in any event.

i. Nutrients

In the past practices of hydrogeology very little attention has been paid to examine the nutrients in ground waters. There are two primary reasons for this. First, it has not traditionally been considered that microbial activities in ground waters are a significant event outside of the hygiene-risk that can easily be monitored by determining the presence or absence of coliform bacteria. Second, the product waters commonly used for testing have passed through a biological interface in and around the borehole that reduces the nutrient loading in the water. Often nutrient levels only rise in the product water being sampled when the biofouling has matured and sloughing is occurring.

The three principal elements usually associated with biofouling events are carbon (organic), nitrogen (organic and inorganic) and phosphorus (inorganic and organic). The ratio of these three elements is often viewed as critical to the generation of a biomass and hence, a possible biofouling event. Generally, it is the phosphorus that becomes the limiting nutrient to allow growth and activity to continue.

Most commonly, it is considered organic carbon, inorganic (as nitrate, nitrite and ammonium) and organic (as proteins) nitrogen and inorganic (phosphate and polyphosphates) phosphates which are the most common sources of these elements in a biofouling. The ratio to optimally allow growth has a C:N:P (carbon:nitrogen:phosphorus) ratio of 100:1:0.25. All three elements have to be in forms potentially available to the microorganisms to be included in the ratio. While the carbon ratio may fluctuate quite considerably, the nitrogen: phosphorus ratio may shift the activities of the microorganisms in the biofouling. The optimal N:P ratio is usually between 4 and 8:1. If the ratio is >8:1, then there would be a deficiency in the amount of available phosphorus and this may be restricting the amount of biofouling. If, on the other hand, the ratio is <1:1, the dominant microorganisms in the biofouling may shift towards those able to fixate molecular nitrogen (N_2, dinitrogen). This is an energy expensive alternative and the rate of biofouling may become reduced.

There are a number of significant factors that have delayed the use of the nutrient concentrations in the water as a method for predicting the amount of biofouling. These include the following facts. First, the water sampled is often product water in which many of the nutrients would already have been removed by the down-hole biological interfaces. Second, the difficulty of determining which substances bearing a particular nutritional element are available for growth and biological activities and which cannot be used (recalcitrant). It is, therefore, becoming more common to attempt to determine the status of the biofouling mass itself in terms of its location, mass, volume and composition, rather than attempt to predict the likelihood of such an event occurring. This involves a direct determination of the form and extent of the biological challenge to the integrity of the well system.

> **Nutrient Loadings in Water Wells Samples
> May not reflect the true levels because of:**
>
> - **Accumulations particularly of carbon and phosphorus in the growing plug**
> - **Erratic sloughing of the matured plug causing sudden surges in the carbon, phosphorus and nitrogen**
> - **Localized denitrification may cause reduced nitrogen levels (under anaerobic conditions)**
> - **Nitrate levels may be exaggerated under aerobic conditions by radical nitrification in the plug**
>
> **Care should be taken to take a sequence of samples while pumping to minimize these effects. For economy, the samples can be combined into a single composite sample for analysis.**

Figure Five

3. Biological Challenge Determination

Traditional techniques to evaluate the potential for a biological challenge to a well system has centered on the adaptation of microbiological techniques developed by the medical industry. Two major tests which have traditionally been used are variations of the coliform test (to determine the hygiene-risk factor) and selective agar spreadplate techniques (to quantify the population size in the water). While the coliform tests have been shown to be applicable to waters and wastewaters, the various spreadplate techniques have generally underestimated the population. This may, in part, be due to the bacteria present in waters being adapted to a relative stable regime in which the nutrient supply is generally in a critical (minimalist) status and the incumbent microbes have to cooperate within consortia (including several strains of different bacteria) in order to survive. Taking these nutritionally stressed microbes and placing them in a dispersed state on an agar-based medium that is relatively rich (even as bacteriological grade agar alone) in organic and phosphatic compounds can cause trauma in the bacterial community. Furthermore, the organisms have essentially to "mine" the water from the agar, a task that some microorganisms may not be able to perform. To compensate for these stresses, very prolonged growing (incubation) periods can be used of 14, 28 or even 42 days to allow time for the bacteria to recover and grow to the size of visible (and therefore countable) colonies.

As is the case with the nutrients, the water sample taken from the product water may not, in fact, carry many or any of the bacteria that may be growing in the biofouled zones within slimes, biofilms or encrustations, etc. A water which has passed through a stable and healthy biological interface may not "pick up" any of the microorganisms that are active there. When the waters do "pick up" some of these microbes, they may be in particulate (biocolloidal) structures and not easily dispersed to allow spreadplate analysis. There have been many occasions when performing agar spreadplate analysis on waters from biofouled water wells, when the biocolloids actually pass down the dilution series before actually being entrapped and, therefore, countable on an agar plate. Traditionally, the lack of correlation between the microbiological test procedures, based in part as they are medical techniques, failed to give adequate correlations which would be acceptable to the engineering profession. Consequently, the role of microorganisms in the biofouling of engineering structures has gone under-reported and generally not recognized as a significant management concern.

D. RECENT TECHNOLOGICAL ADVANCES

Perhaps the major breakthroughs in microbiology over the last twenty years relate to the growing recognition that, in the natural world, microorganisms commonly grow attached to surfaces or each other in biofilms. This further involves different strains of bacteria cooperating with each other within the biofilms and form consortia that may become very stable entities. An interesting example of the latter event is the consortium that forms in the black plug layer in high-sand content turfgrass golf greens. Here, lateral slime forms which can directly compete with the turfgrass for water, oxygen, nutrients and the voids in the sand. This consortium routinely includes *Erwinia carotovora* (a bacteria which causes soft rot in plant roots) and *Bacillus thuringiensis* (the bacteria which can kill insect larvae). The consortium, therefore, includes a specific bacterium which can initiate the rotting process in the roots of the turfgrass, and another bacterium which can kill at least some of the predators that may attempt to feed on the plug layer. Most members of the black plug layer are able to generate slime (scientifically referred to as extracellular polymeric substances, EPS) and this hold is to protect and secure the consortium. In water well biofouling, the consortia that form now appear to be much more complex in structure and often exist within a number of distinct biozones.

Recent technological advances have resulted from a recognition that the biofouling is not simple bacterial infestation but includes complex structured consortia which function within related biozones. Past practices of single form chemical treatment have failed to recognize the ability of these organisms to recover from such trauma and rapidly reoccupy the spaces vacated due to the chemical treatment. Current technologies being field tested and commercialized today recognize the complexity of the biofouling and take a more strategic approach to controlling rather than destroying the causant microorganisms of the biofouling. In medical practice, it is presently generally believed that all infections are the result of a successful infestation of the patient by one strain of microorganism (i.e., the pathogen). Applying an antibiotic known to control (and preferably kill) that pathogen is a common strategy in practice. In biofouling events, the infestation of the patient (e.g., the well) involves a complex of consortia each with a different vulnerability to chemical treatment. Strategies in the last two decades have adapted to the concept that a radical or multiple challenge treatment can be more efficient.

Radical treatments range from the generation of a temperature gradient either upwards (to kill the incumbent organisms in the biofouling) or downwards (to freeze the biofouling at which time it often detaches); to the application of more radical chemical treatments with elevated dosages and/or prolonged exposure times. Multiple challenges can be created by applying, sequentially or concurrently, a combination of treatments in a manner that is synergistic to the destruction of the biofouling. Often, such strategies include the use of dispersants, pH modifiers and disinfectant agents in a manner that maximizes the recovery of the biofouled well to optimal conditions. One of the most recent treatment technologies (Blended Chemical Heat Treatment, BCHT™) which is patented involves a triphasic treatment.

> **Rehabilitating a Plugged Water Well should involve:**
>
> - **SHOCK** – to traumatize the microorganisms forming the plug
> - **DISRUPT** – to break up the structures forming the plug
> - **DISPERSE** – to clean off the surfaces onto which the plug had been growing (remember that cleaning must also include getting rid of any debris)

Figure Six

The three phases are: (1) SHOCK, (2) DISRUPT, and (3) DISPERSE. In other words, the incumbents in the biofouled zone are first of all shocked (i.e., traumatized); now the structures within the biofouled zone (e.g., the biofilms) are disrupted; and finally the disrupted biofouling is dispersed away from the zone of concern.

As the ability to conduct adequate determinations of the nature of the biofouling and its site, it may be expected that there will be a growing acceptance of the need to manage biofouling as a controllable problem. To do this, adequate diagnostic techniques have to be used, appropriate rehabilitation performed (to recover the production capacity of the well), and a preventative maintenance program needs to be established to ensure that the useful lifespan of the well can be extended. These goals are very much in harmony with the shifting mindset of society from disposability to reusability, and from environmental negligence to environmental responsibility.

III. THEORETICAL APPRAISAL: CAUSES OF BIOFOULING

One of the major disadvantages of determining the cause of biofouling is that it is often very difficult to determine the site of the biofouling (since it is often out of sight and out of mind). Frequently, it is more convenient to attribute the perceived malfunctioning in the process or system to some physical and/or chemical anomaly rather than to address the question from a biofouling perspective.

The science of biofouling is still very young, and because of the inevitable lack of recognition generally attributed to a young science, much of the information is anecdotally (by word of mouth and generalist articles) and experientially (by the descriptions of the observer) presented. Like any other new sub-discipline there is the inevitable generation of terms to ascribe the various events and processes that become a part of the understanding. This chapter is presented in five parts which are related in some manner to gaining a better understanding of that "beast" which performs feats of biofouling sometimes covertly (not obviously detectable) and sometimes very obviously (slimes oozing out of valves, equipment corroding).

It is necessary to understand biofouling here is a fuller description of the various forms of biofouling (i.e., know the "beast" and in the knowing then shall its weaknesses become apparent). Like any other living complex, biofouling passes through a number of stages in its maturation to senility.

The forms of these growth cycles can create in the reader's mind an understanding of the dynamics of a biofouling event. It becomes essential to be able to utilize as simple and reliable as possible indicators reflecting whether a biofouling is present or absent and, if present, what stage the biofouling is in. To distill the inevitable mass of information into some form of logical sense, there are last aspects of critical importance that need to be addressed. These are: (1) confirmation of the occurrence, size, state and site of the biofouling; and (2) the generation of a preventative maintenance program which will allow a satisfactory monitoring of the status of the biofouling after control measures have been instigated. In other words, know thou the "beast" and where it does hide; and when the "beast" is controlled, then guard thyself against its return!

A. MAJOR SYMPTOMS OF BIOFOULING

Biofouling is, in some ways, a very new phenomenon and yet it has existed in nature throughout time as natural events. The fact that these events can often occur naturally leads people to accept the effects as a "normal" act of nature and a part of the normal risks that have to be accepted. Biofouling is, however, a complex event and yet each event is, in some senses, unique. While the effects (or symptoms) may be definable the cause may be complex and difficult to determine. As J.H. Woodger wrote in "Biological Principles" published by Routledge and Kegan Paul, London in 1929, this was summarized by the statement that: "All the wood stands in a mist of green and nothing perfect". In biofouling even the "mist" is difficult to see and indeed the presence of "nothing" can never be considered as perfect states (or rather states of unawareness)!

Frequently, biofouling events have been considered to be primarily physical and/or chemical events because the effects were often readily interpretable by those techniques. For example, a loss in flow from a well was simply a result of silting or encrustation (i.e., physical blockage and chemical plugging). The corrosion of an iron pipe may be considered to be simply an electrolytic corrosion event which is primarily physical in form. There is a rapidly growing awareness that biofouling can no longer be treated as simply a necessary but unpredictable risk. With the advent of improved microbiological methodologies, biofouling is becoming definable and measurable. As there is a growing need to extend the life expectancies of installations to make them environmentally sustainable, there needs to be improved methods of control. This then translates into economic verification as a controllable problem which, when addressed, allows economic extensions to the useful lifespan of the installation.

Often the "cause and effect" of biofouling are addressed backwards. This is done because very often it is the symptoms of biofouling (i.e., effects) which become evident and set in motion a rehabilitation followed by (hopefully) a preventative maintenance program to control the problem. The "cause" of biofouling must be approached first.

B. PLUG FORMATION

Plug can be defined as "an encumbrance or impediment," while plugging may be considered to "become obstructed, especially by the accumulation of a glutinous mass." This definition covers many of the primary aspects of plugging in biofouled water wells. First, there is a glutinous mass that could be related to the formation of slime from coalescent biofilms. Second, there is an encumbrance, impediment (to flow) which is caused as the plug fills the void spaces in the porous media and reduces water flow. Third, there is an accumulation of material within the plug indicating that it is growing and therefore likely to become more obstructive. Clearly, from the brief definition, many aspects of the growth can be projected. There are two major features of concern: accumulation and obstruction.

A plug can be considered to have four parts that dominate at different times in the growth cycle. These are:

1) The living microbial cells form the "heart" of the plug. It is thought they may not occupy a very large part of the volume.
2) The EPS forms a large part of the volume of the plug. It forms the bulk of the glutinous mass and holds "bound" water along with other accumulates.
3) Amorphous chemical complexes formed within, or attached to, the plug.
4) Crystalline chemical complexes formed directly in the plug or via the amorphous forms. Crystalline complexes often arise at interfaces between the plug and either solid surfaces, the free flowing water, or the atmosphere. These chemicals may be dominated in particular by carbonate, oxide and hydroxide salts.

The mechanism of growth causes a range of effects that relate to the relative state of the plug. During the growth of a plug, the above parts change (Table One) in their dominance of one to the other.

Table One
The Relative Shifts in Dominance during Plug Formation

	1 Viable cells	2 EPS slime	3 Amorphous	4 Crystalline
Attachment	+++	+	−	−
Early growth	+	+++	−	−
Stabilization	+	++	+	−
Cyclic growth	+	++	++	+
Encrustation	+	+	++	+++
Occlusion	+	++	+++	++

Note that the number of + signs represents the degrees of dominance and the negative (−) sign reflects absence.

As the plug forms it goes through a number of stages of maturation. Initially, there has to be attachment of microorganisms to the surfaces within the porous medium or non-porous materials. This attachment is achieved by the viable cells "throwing down" EPS to anchor the cells to the surfaces.

Early growth involves the attached (sessile) microbes now producing copious EPS that occupies a very considerable amount of the void volume if porous media are being infested. The net effect of this is that there is a loss in porosity and the water flow may be severely impaired. This is a temporary phenomenon that is sometimes seen to influence initial production from new wells before they have been "developed."

Very rapidly, the initial slime formation condenses and stabilizes with a residual very thin growth remaining over the infested surfaces. Flows now return and stabilize. In laboratory microcosms, the flow from a newly stabilized plug formation can exceed even the flow from through the porous medium in the pristine (non-fouled) state. There now begins a period of cyclic growth in which there is a pulse-like increase in the volume of the plug as it occupies more and more of the void space. From the laboratory and some field experiences, the pattern always appears to be the same: (1) expansion of the volume; (2) sloughing; and (3) stabilization. In production wells, this pulsing may be seen by regular shifting in the drawdown of the well, depending upon the point in the life cycle that the plug is in. During this cycle there is a gradual increase in both the amorphous and crystallizing forms occurring within and around the plug. A process of bio-accumulation of recalcitrant chemicals is underway, which means that there is a shift in the dominant components away from the viable cell (1) and the EPS (2) towards the amorphous (3) and the crystalline (4) components. Both the amorphous and crystalline deposits are essentially recalcitrant in that the residual microbial population may not be able to utilize these chemicals for any purpose. However, the energetics of the system remains to be established.

When encrustations are produced, these may be dominated by a crystalline form of carbonate and oxides. The surfaces become smooth and hardened, and the microbial population and EPS content shrinks very considerably. Generally, an established stable encrustation does not radically affect flow or water quality to a variable extent. Any changes that do occur are relatively slow in appearing. When occlusion of the void spaces occurs, there will still be a significant microbial population with a significant EPS present. Much of the chemicals may remain in the amorphous form, but the occlusion can cause total loss in flow through the infested porous media.

Plug formation may, therefore, have a number of sequential impacts on the water being drawn from the wells as a result of the different stages in the plug formation. The symptoms will change with the age of the plugging and the rate at which it is generating.

C. SEQUENCE OF SYMPTOMS IN PLUG FORMATION

The following sequences form the normal events which occur during the development of a plugging condition:

1) Short unstable period of production begins as the plug goes through the early stages of growth. This is associated with the high (but short-lived) void volume occupancy that occurs during early colonization.
2) Good production of water continues at capacity with the product water being of a higher quality than the ground water feeding the well. This is because the biological interface being established as a part of the plug around the well is removing nutrients and various metals (e.g., iron, manganese, zinc, aluminum, copper, etc.) from the water as it moves across the redox front. The nutrients are utilized by the plug formation for growth while the metals and other recalcitrant material bioaccumulate.
3) Water production becomes variable. The drawdown against a stable pump rate may vary on a day-to-day basis and reflects the status of biofilms within the plug as they pass through growth, slough and stabilization phases. Water quality may become more variable; declining during the sloughing, and recovering and stabilizing during the other phases. This variability in the water quality sometimes creates an attitude of uncertainty in the interpretation of the data obtained.
4) There now develops a general but erratic loss in both water production rates and the quality of the water. Frequently, the TSS in the water increases along with color. This is a reflection of the radical but often random sloughing of the plug formations. These symptoms are the result of an increasing amount of sloughed particulates that still retain the bio-accumulates. Some of these accumulates may also be present and dissolve (and color) into the water itself. Water production is affected by the reduction of porosity created by the plug now filling a significant percentage (usually >40%) of the void volume. This may be reflected in increasing drawdown within the well and an increased probability of over pumping causing the drawdown to cause the pump to "suck-air".
5) The terminal state is reached when the production capacity of the well is so impaired that it is not practicable to rely on the well as a source of water. The plug has essentially occluded

(plugged up) the formation and is acting as an unstable structure. Such instability leads to the product water quality degenerating still further and becoming unacceptable even if the flows were adequate. Such wells now require radical rehabilitation in order to disperse the plug and return the well to an acceptable production. Failure to rehabilitate the well for economic or other considerations leads to the abandonment of the facility.

There are reports of abandoned plugged water wells returning to an acceptable production after they have been left essentially untended for a period of years. These "spontaneous" recoveries from plugging may be a reflection of the changing environment that occurs in and around the well after it has been abandoned. The well reverts to become simply a biofouled part of the ground water system. As a result, the environmental characteristics of the habitat around the well changes as the redox front either dissipates or moves. In either event, the microorganisms still occupying the plug formation will tend to migrate towards a more acceptable environment (e.g., another redox front somewhere else). This act of migration can destabilize the plug structure and cause a gradual dispersal of the elements concentrated within the plug. Over time such traumatized plugs may become dispersed completely so that the well can, at least theoretically, be brought back into production after suitable rehabilitation.

As the cost of operating water wells increases for various reasons, this option of rehabilitating even totally plugged installations becomes a more tempting option that is only now beginning to be appreciated. Various techniques discussed in later chapters can be applied to such "last resort" rehabilitations.

D. RECALCITRANT CHEMICAL ACCUMULATES

Plug formations involve a system of bioaccumulation which may be active (microorganisms utilize some element of the accumulation for a specific purpose) or passive (the deposition is purely a physical and/or chemical event not involving any microbial activity). The chemical accumulates fall under two general headings: amorphous and crystalline. Amorphous deposits have an indefinite shape and structure while crystalline deposits usually have a more defined structure and edges sometimes with evident internal structures. These accumulates are recalcitrant (not immediately utilizable by the microbes within the plug) but may become modified (e.g., to a different form or dissolved) by subsequent action which may or may not involve microbial agents.

There is a very limited knowledge of the role of microorganisms in the precise mechanisms involved in bioaccumulation. Of the crystalline deposits, carbonates, oxides, hydroxides and oxyhydroxides may dominate. Calcite ($CaCO_3$) may be a major carbonate along with some siderite ($FeCO_3$) where iron is being bioaccumulated. Other iron compounds that may be present include hematite (Fe_2O_3), and geothite ($FeO(OH)$). Non-crystalline (amorphous) minerals commonly found may include ferrihydrite ($Fe_5HO_8.4H_2O$). Both crystalline and amorphous forms of chemicals are often complex involving a range of substances involving other elements and complexes.

The nature of the chemistry of these events is that each plug formation may have a unique composition and thus, will have a different nature in its growth and control. As more is discovered about the nature of plug formation a better understanding should be generated as to their control.

Another factor affecting the characterization of a plug is the fact that the chemicals being accumulated do not do so in a random manner. There is awareness that the cations (e.g., metal ions) accumulate in a sequence as the water flows over the plug. Wells in which the plug has been dispersed and surged out of the well sometimes exhibit changes in the ratios of the metallic elements as the dispersed plug is being pumped from the well. Iron, for example, tends to accumulate on the oxidative side of the redox front. As a result, much of the bulk of the iron in the ground water is removed early in the pumping procedure. Manganese, on the other hand, tends to accumulate deeper into the plug formation under more reductive conditions. This differentiation of the iron and manganese into different sites of accumulation may relate to the polarizability, number of outer shell electrons, and the degree of symmetry in the cation.

Where organic compounds are involved in the accumulation process, there may be a form of bonding between the metal and the ligand (organic site of the bonding). There are two mechanisms in which this bonding can occur. First, the bond may be formed as ion pairs in which both the metal and the ligand retain hydration (bound water) sphere. Second, a coordination complex may form in which the ligand is immediately adjacent to the metal and some degree of electron donation occurs between the metal and the ligand. Sometimes these complexes can involve multiple ligands and form a chelate complex.

The nature of these very complex structures renders any control strategy that much more challenging to implement. Not only that, the effect of dispersing and pumping out a plug formation means that the product flow will contain the various chemicals accumulated within the plug. One irony of this is that the plug may have been "filtering" the ground water for a number of years. This would provide a better product (post-diluvial) water until the plug begins to slough.

E. COMPROMISED WATER QUALITY

In general, the water quality for a given well is established after the well has been "developed" and gone into service. This act of development is a period during which the well stabilizes. Stabilization may also be viewed as occurring at the beginning of the third phase of the biofouling after compression of the plug. Compromise occurs as the water quality begins to degenerate as the plug matures. Symptoms will vary from well to well, but the general features that may be expected are listed below by each parameter.

1. Iron

There is a phased effect in which there would initially be an erratic rise in the amount of soluble iron in the water which may color the water to a yellow or orange hue. However, there may not be any significant increase in turbidity. This would be because much of the iron may be retained within small biocolloidal particles which would not influence the turbidity so much. There may be some variability in the total iron recorded and this would reflect the stability in the plug formation (e.g., the degree of surface sloughing occurring). Repeat samples over a period of a few days may show significant differences between each other, since the plug may be in a stable-unstable cycle within the maturation period.

As the plug continues to mature, the sloughing events may become magnified leading to the secondary symptoms. These include a degenerating increase in turbidity, higher total iron content, larger particles and a high TSS. This may often be accompanied by significant (>15%) drops in the production capacity of the well due to occlusion. As this happens there may be a very significant precipitation of these particles in the well water column leading to the buildup of a sediment which may re-suspend when the well becomes active. Very often, the early flow from an infested well will be very high in particulate material and iron and should be diverted to waste to avoid seriously challenging the downstream systems and processes.

2. Total Organic Carbon

This is a prime source for the growth of heterotrophic microorganisms, since this contains both the source of carbon and often the source of energy for the microorganisms. Because of this, the TOC in the water may be lowered as it passes through the plug formations around the well. In laboratory mesocosms it is not uncommon to find greater than 85% of the TOC becoming bound within biofilms (and, therefore, removed from the water). Generally, TOC values will remain low until there is a destabilization and massive sloughing from the plug formations. When this happens, the TOC values may become elevated by as much as an order of magnitude. These elevations may be unstable and a reflection of the degree of sloughing that is occurring along with any re-suspension of materials that have precipitated in the bore hole and other "dead" ends within the system. TOC in the product water cannot, by itself, be used to diagnose the likelihood of a biofouling event and usually reflects the occurrence of such an event after there have been other, and often more obvious, symptoms.

3. Total Nitrogen

Nitrogen is very often not estimated in water wells beyond nitrate that forms a hygiene-risk to infants when the concentration exceeds 10 ppm. The form in which the nitrogen occurs in the water can be used to project with some accuracy the form of biological action that is occurring in the upstream biological interfaces. Nitrogen is commonly estimated using the following parameters:

N3	Nitrate nitrogen	- $N-NO_3$
N2	Nitrite nitrogen	- $N-NO_2$
NH	Ammonium nitrogen	- $N-NH_4$
NO	Kjeldahl nitrogen	- organic nitrogen + NH

The ratio for these different forms of nitrogen can give an indication of the status of the biological systems in the upstream interfaces. Given that the total nitrogen (TN) is:

$$TN = NO + N3 + N2$$

A number of extrapolations can be made and these are addressed below.

>0.2 = NH/NO: There is a significant amount of ammonium relative to the total Kjeldahl nitrogen, which would suggest that the water contains non-oxidized (anaerobic) degradation products, and any plug

formation may be dispersed since no redox front has become established.

>0.4 = N3/NO: Nitrification is likely to have been occurring extensively upstream in an oxidative environment, possibly through a shallow aerobic polluted recharge. Frequently, such pollutants may be relatable to sewerage or septic tank wastes that are degrading aerobically. The nitrate concentration may be transient (and variable), since many microorganisms are able to use nitrate as an alternate respiratory agent to oxygen.

>0.6 = N2/N3: Denitrification is occurring in which the nitrates are now (anaerobically) being reduced via nitrite to nitrogen (complete denitrification). Usually, this occurs on the reductive side of the redox front, the nitrate concentration may be variable and generally decline as the denitrification continues.

>0.8 = NO/(N3 + N2 + NH): The amount of organic nitrogen considerably exceeds the sum of the inorganic nitrogen that would imply that a sloughing (or growth) is dominating the nitrogen mass balance in the water. Active biofouling (such as plug formation) may be suspected to be present upstream in this event.

4. Total Phosphorus

Phosphorus in biological systems is akin to the battery in an automobile. It provides the energy storage system (as ADP and ATP) which drives the biological systems. It is, therefore, a very essential and sought after element (rather similar to the human desire for gold, biological systems desire phosphorus). The net effect of this essential need is that phosphorus is "hoarded" by biological systems and may not, therefore, be as readily released into the water in the soluble form. In the cell, phosphorus may occur as orthophosphate (the currency), polyphosphate (the reserve) and metabolic phosphorus (the energy driver). In water, phosphorus is normally found in four states:

- SIP Soluble Inorganic Phosphorus
- PIP Particulate Inorganic Phosphorus
- POP Particulate Organic Phosphorus
- SOP Soluble Organic Phosphorus

These, together, form the total phosphorus (TP). The ratios of these fractions of the phosphorus pool can be used to diagnose some biofouling problems.

>0.3 = SIP/TP: A dysfunctional system in which there is some level of trauma for whatever reason is allowing the SIP (usually as orthophosphate) to pass through the upstream environment without being taken up by the biological interfaces. The trauma may be created by some radical shift in the conditions resulting from such events as a disinfection treatment or a sudden and dramatic change in the environmental conditions.

>0.5 = PIP/TP: Much of the phosphorus may be in a recalcitrant form not readily utilizable by the upstream biological interfaces. Such phosphorus may not necessarily be relatable to the amount of biofouling that can be expected. It should be remembered that this ratio may be distorted upwards by the retention of the other forms of phosphorus (SIP, SOP and POP) within the zone(s) of biofouling.

>0.2 = SOP/POP: An inordinately high SOP to POP ratio would indicate that there was either a disintegration of the biological interface with the releases of SOP from the cells that constitute the POP. This could happen where there has been a successful dispersion of a plug (for example).

>0.85 = POP/TP: Much of the phosphorus is probably bound within biocolloids and other particles. Much of this phosphorus may be in the form of reserve polyphosphates; the organisms within the particulate mass are probably still viable and have not been disrupted.

An idealized ratio for the SIP: PIP: POP: SOP in a very active planktonic system may be expected to be in the range of 0.1 - 0.2: <0.05 : >0.8 : <0.1 respectively. However, in a ground water situation, the "site of growth (eutrophication)" may be some distance from the sampling point; and so the ratio in the sampled water will, most likely, be distorted away from the optimal towards that generated by the sloughed and recalcitrant chemicals passing through the system.

Phosphorus is commonly perceived as the major nutrient limiting growth. One of the reasons for this is that the phosphorus is such a desirable (and storable) nutrient. Very often, analyses of product waters reveal very low levels of phosphorus, and the natural conclusion is that phosphorus must, therefore, be the limiting nutrient for growth. In reality much of the phosphorus could be tied up by the biomass (concentrated at the redox fringe) as polyphosphates. To truly determine the plugging potential, the mass balance for phosphorus within the likely zone of plugging would need to be determined.

5. Redox Potential

It has already been stated that the biological (and relatable plugging) activity will focus at the redox front as the water moves from a reductive to an oxidative state. Product water is commonly sampled from the oxidative side of the redox front and may, therefore, not be very useful in determining the position of the front itself. However, if there is a low redox value (e.g., +50 - 0.00 millivolts) then the site of biofouling is probably closer to the sampling site. A negative redox value (-0.00 - -200 millivolts) may indicate that the major biofouling could be occurring downstream from the sampling point.

6. Temperature

It has now generally been believed that a temperature gradient can be created in a well that is actively undergoing a biofouling. Recent studies have revealed that minor gradients can be created which are measurable and do focus on the sites of biofouling. Generally, the temperatures found in most water wells fall within that broad band of temperatures (6 to 40°C) where a wide spectrum of biological activity is potentially able to be active.

7. Microbiological

The irony of determining the presence of microbiotic agents within product water is that most of the biological activity will be at the plug formation and other sites related to the redox front. The attached (sessile) microbes growing within the fouled zone itself will likely dominate activities. Consequently, the product may contain anomalous (commonly low) presence of microorganisms compared to the true amount of activity that is occurring. Historically these observations, together with the absence of coliform bacteria, have led to an underestimation of the role microorganisms may play in the fouling process.

F. PRODUCTION FAILURES

Given that the water is commonly sampled from the well water column or downstream, the water quality has already been compromised by the upstream biofouling, the product water quality and the flow may degenerate very slowly if the production demands are close to the wells designed capacity. If the well is being operated below capacity, then the impairment of flow by plugging may be a relatively dramatic event.

G. APPRAISAL OF THE CAUSES OF BIOFOULING

One of the major disadvantages in determining the cause of plugging is that it is often very difficult to determine the site of the biofouling (since it is often out of sight and out of mind). Frequently it is more convenient to attribute the perceived malfunctioning in the process or system to some physical and/or chemical anomaly rather than to address the question from a biofouling perspective. The science of biofouling is still very young, and because of the inevitable lack of recognition generally attributed to a young science, much of the information is anecdotally (by word of mouth and generalist articles) and experientially (by the descriptions of the observer) presented. Like any other new sub-discipline, there is the inevitable generation of terms to ascribe the various events and processes that become a part of the understanding.

Problems in the Determination of Plugging in a Well

- **Lack of records on the well's performance**
- **Operating the well at much lower levels than it is capable of**
- **Symptoms occur slowly and are not recognized**
- **Uncertainty in the effectiveness of any selected treatment**
- **Failure to recognize the economic cost benefits of both rehabilitation and preventative maintenance**

Figure Seven

It becomes essential to be able to utilize as simple and reliable as possible indicators reflecting whether a plugging is present or absent and, if present, what stage the biofouling is in. To distill the inevitable mass of information into some form of logical sense, there is also a discussion of the appropriate selection of parameters to be used to monitor the biofouling. Two aspects of critical importance are addressed. These are: (1) confirmation of the occurrence, size, state and site of the plugging; and (2) the generation of a preventative maintenance program which will allow a satisfactory monitoring of the status of the biofouling after control measures have been successfully instigated. In other words, know thou the "beast" and where it does hide; and when the "beast" is controlled then guard thyself against its return.

IV. CHRONOLOGICAL SEQUENCES

A. WATER PRODUCTION

Initially, there will be a short unstable period of production as the plug goes through the early stages of growth. This is associated with the high (but short-lived) void volume occupancy that occurs during early colonization.

After this, there may be a period of good production of water at capacity with the product water being of a higher quality than the ground water feeding the well. This is because the biological interface being established as a part of the plug around the well is removing nutrients and various metals (e.g., iron, manganese, zinc, aluminum, copper etc.) from the water as it moves across the redox front. The nutrients are utilized by the plug formation for growth while the metals and other recalcitrant material simply bioaccumulate.

In the next stage, water production becomes variable. The drawdown against a stable pump rate may vary on a day-to-day basis and essentially reflects the status of biofilms within the plug as they pass through the growth, slough and stabilization phases. Water quality may become more variable, declining during the sloughing, and recovering and stabilizing during the other phases. This variability in the water quality sometimes creates an attitude of uncertainty in the interpretation of the data obtained.

Gradually, there now develops a general, but erratic, loss in both water production rates and the quality of the water. Frequently, the TSS in the water increases along with color. This is a reflection of the radical but often random sloughing of the plug formations. These symptoms are the result of an increasing amount of sloughed particulates that still retain bioaccumulates. Some of these accumulates may also be present and dissolve (to color) the water itself. Water production is affected by the reduction of porosity created by the plug now filling a significant percentage (usually >40%) of the void volume in the infested zone. This may be reflected in extended drawdown within the well and an increased probability of over pumping causing the pump to "suck-air".

The terminal state is reached when the specific capacity of the well is so impaired that it is no longer practicable to rely on the well as a source of water or to provide hydraulic relief. The plug has essentially occluded (plugged up) the formation and is acting as an unstable form of "dam" structure. Such instability leads to the product water quality degenerating still further and becoming unacceptable even if the flows become adequate. Such wells now require radical rehabilitation in order to disperse the plug and return the well to an acceptable production. Failure to rehabilitate the well for economic or other considerations leads to its abandonment.

Given that the water is commonly sampled from the well water column or downstream, the water quality has already been compromised by the upstream biofouling. Product water quality and flow may be expected to degenerate very slowly if the production demands are considerably below the well's designed capacity. Thus, if the well is being operated below capacity, then the impairment of flow by plugging may be a relatively dramatic event.

B. BIOFOULING PROCESS

There are a number of stages that commonly occur during the process of biofouling in water wells. These stages are affected by a number of major factors. These include the water quality and production rate, the nature and porosity of the media around the well, the design and construction of the well and the degree of preventative maintenance which is being applied during the operation of the well. There remains, however, a sequence of events that may be commonly expected to occur. These are:

1. Initial Colonization

It is to be expected that the natural (intrinsic) and introduced (extrinsic) microorganisms will be present within the environment affected by the installation of the well. The intrinsic flora will have arisen from the ground water, the unsaturated media above the water table, and from the soil. The extrinsic microorganisms are those introduced to the environment by the introduction of drilling, waters, personnel, muds and various chemicals required for the construction and development of the well. These microorganisms will compete for the available surfaces onto which they attach. Desirable surfaces would be charged, occur at the redox front, sited where turbulence is being generated in the water, are along pathways of nutrient supplies (e.g., organic deposits, higher concentrations of dissolved/suspended organic carbon, phosphate rich zones). The act of attachment involves the microorganisms attaching, reproducing and growing outwards to form a biofilm. As the various microbial biofilms form, they interact and consortial (community) biofilms form. Each of these will contain a number of strains that cooperate within a common biofilm.

2. Primary Void Volume Occupancy

Once a coherent consortial biofilm has formed over the surfaces, there is now competition to occupy the void volume of the porous media. The void volume is that volume within the porous media that can be occupied by water in a saturated condition. In this initial competition between the various consortial biofilms, the volume of the biofilms expands rapidly to reach to 10 to 60% occupancy of the void volume. Transmissivity of water through the formation may be severely impaired at this time but the void volume occupancy reduces, often rapidly, as the biofilms now stabilize.

3. Primary Stabilization

The biofilms now stabilize with small void volume occupancies. Transmissivity returns to pre-attachment levels and the production capacity stabilizes. At this time, the well is usually considered developed and goes on-line as a fully functional unit.

4. Secondary Void Volume Occupancy

There is now a pulse-like growth in the biofilms involving three phases. The three phases in the cyclic growth are: (1) volume acquisition; (2) surface sloughing; and (3) re-stabilization. At the end of each cycle, the void volume occupancy has further increased impairment of the transmissivity of water through the biofouling of the porous media. These cycles can cause fluctuations in the pump rate when pumping is done to a fixed drawdown, and the well is being operated at close to full capacity. During this phase of growth, much of the nutrients are being retained within the biofilms along with any other recalcitrant chemicals (e.g., iron, manganese) which are bio-accumulating. The product (postdiluvial) water coming from the well will show lower chemical concentrations due to this phenomenon, except during the sloughing stage of the cycle. At this time the chemical and particulate loading in the product water may increase dramatically causing the water quality to periodically degenerate. Plug formation shifts during this phase to an encrustation or dense amorphous slime formation.

5. Plugging

Loss of available void volume and the growth of a plug formation eventually lead to significant losses in the capacity of the well. The loss in transmissivity through the biofouled media restricts the amount of water reaching the well. In general practice, this loss in capacity may be observed as the production falls to between 50 and 80% of the original capacity as a result of the plugging. There are two factors that influence this loss in capacity. First, there is the loss in void volume capacity to allow water flow. Second, as the plug forms within a porous medium, a localized plugging may isolate some zones within the formation. This can reach the state in which some of the water becomes "locked" within an encircling plug formation. This severely reduces the transmissivity of water towards the well and can cause radical reductions in the production capacity of the well.

6. Total Plugging

The extensive development of plug formations maturing towards dense amorphous or crystalline forms together with the "locking" of water within the porous media. This leads to a critical loss in the capacity of the infested well and causes sudden, and often dramatic, a drop in the capacity of the well to the point that it is no longer economical and/or desirable to keep the well online. Sudden plugging may be expected to occur when the well is down to between 20 and 60% of the original capacity. When total plugging occurs, the capacity often falls suddenly and the well may even become essentially "dry" (producing water in such small quantities that its use is not practicable).

Unfortunately, many wells are not utilized to their full capacity. One result of this is the difficulty for the operator of the well to appreciate that a plugging is occurring until the loss in capacity becomes greater than the production demanded of the well. By then the well may be so severely plugged that rehabilitation of the well becomes much more challenging to achieve. In the next section, emphasis will be directed at the various parameters that can be used to determine the degree of plugging which has occurred and the site and extent of the plug(s) that may be present.

C. CRITICAL INDICATORS: THE STAGES OF BIOFOULING

As the need to maintain a functioning water well of any type becomes more critical, there is going to have to be: (1) an appreciation of the inevitability of some level of biofouling; and (2) an understanding that the product water from the well will not necessarily reflect the risk or level of biofouling. The challenge is, therefore, to develop a system, which would make allowances for these uncertainties in the prediction and monitoring of the plugging risk index. Indicators include chemical, physical and biological parameters that can be used to determine the theoretical position of the plug formations within the porous media surrounding the well at risk.

D. CHEMICAL INDICATORS REFLECTING BIOFOULING

Since it is taken as almost certain that the water sample being subjected to analysis and interpretation is product water sampled from the well water column or downstream, the comments below relate to the determination of the plugging risk index based upon the data so gathered.

The classical techniques of determining the Biological Oxygen Demand as a measure of the (downstream) requirements for oxygen that are likely to be required are not so relevant when attempting to determine (upstream) biological activity potentials. The irony is that the analyses will tend, in the early stages of plugging, to indicate that the product water is of better quality than it really is (i.e., prior to passing through the plugging zones around the well). Chemically, the plugging becomes assessable once the plug enters the phases of cycling in which the biofilms pass through growth, sloughing and stabilization phases in a repeatable manner. One effect of this is that the water will periodically contain sloughing elements from the biofilm that will materially change the chemistry of the water. This would result in the water possessing a greater variability in the targeted chemical parameters that are associated with the sloughing. Parameters likely to be included in such variability are:

> Total Organic Carbon
> Kjeldahl Nitrogen
> Total Phosphorus
> Total Iron
> Total Manganese

During the stabilization and growth phases, the parameters will be relatively stable in concentration. Once sloughing commences, all of these parameters would increase in the water to a variable extent. Much of the increase in the chemical parameter concentration would be due to the increases in particulate loading (inputs from the sloughing of the plug) rather than in the dissolved fraction. This increase may be ascertained by filtering the water through a 0.45micron membrane filter by gravity flow (without suction). For a given total concentration of a chemical (C_t), the dissolved fraction (C_d) would pass through the filter while the particulate fraction (C_p) would be retained on the filter. Given that the chemical may be differentiated into these two fractions:

$$C_t = C_d + C_p$$

It may be expected that in a case where sloughing contributes only to the C_p fraction, the following characteristics may be expected to occur: C_d would remain relatively constant and the C_p would increase. This would be relevant to the baseline data obtained when the plug is in the stabilization and growth phases of the cycle. Unfortunately, it is not practicable to be able to monitor and project the periodicity of the triphasic cycle. In laboratory and field experiences, this cycle has been seen to occur in periods ranging from days to weeks in length. This makes the use of chemical data difficult to exploit in the prediction of a plugging event at this stage. Unfortunately, the occurrence of unusually high variations in chemical data on some samples may be considered by the interpreter to represent anomalous events. When this happens, their true role as early indicators of a plugging can be overlooked. Additionally, very often the chemistry of a water sample is restricted to only the filtered sample, that is essentially the C_d, which may not shift much during the early phases of plug formation.

Once the plug formation passes beyond the triphasic stage of the growth cycle towards a full maturation and plugging, the chemical parameters now degenerate with permanent increases in the sloughing particulate mass increases and carries with it a greater mass of bio-accumulated chemicals. As the plug matures, the particulate material may become denser and contain higher concentrations of iron and manganese while the relative organic carbon, nitrogen and phosphorus loadings decrease. It is not uncommon for the sloughing particulates from a total plug to contain up to 10,000 to 20,000 ppm (dry weight) of iron in the ferric form. Once this has happened, there is a high probability of total plugging occurring at various points throughout the biofouled zones in and around the well. This has now almost totally prevented flow into the well. This "locked" water may be encircled by plug formations, and so may also be found in zones within the infested porous media. Treatment now becomes much more challenging due to the occlusive nature of the matured plugs. From the chemical standpoint, instability in the selected parameters coupled with a general, but erratic, rise particularly in the suspended solids (particles) can form a warning system that the system is becoming badly plugged and likely to fail dramatically.

E. PHYSICAL FACTORS INFLUENCED BY BIOFOULING

From the previous section, it can be seen that one factor which can be used as a "marker" for biofouling and plug formation is the

development of a higher and higher particulate loading. This is most easily determined as the Total Suspended Solids (TSS) and is usually measured in ppm. The biofilms in the plugged zones slough so higher TSS values may be recorded in the water samples. One of the most satisfactory ways to measure TSS is by the use of laser driven particle sizing and enumeration. This technique allows the size of each particle to be determined and the volume computed together with an enumeration of the total number of particles. To do this, a pulsed laser light is directed through the water sample and the interference in the regularity of the pulses is used to determine the particle characteristics. From the data obtained, the laser driven particle counting is able to generate the following items of data:
> TSS
> Mean particle size (microns)
> Standard Deviation in particle size
> Percentage Volume Distribution in the various allocated sizes
> Combined Surface Area
> Wet Weight (by predetermined density allocation)

This information can overwhelm the user, unless a sense of the relative importance of the data is established. The assessment of the presence and status of a plug formation can involve the following parameters.

1. TSS

All of the particles that interfere with the laser beam are recorded. These range in size from less than half a micron (less than most vegetative bacterial cells) to greater than 100 microns. When there is a sloughing event from a biofouled zone, the TSS will automatically increase as the particles formed by the sloughing enter the water. These particles may vary in concentration as the water moves through different formations and "flushes" away the sloughed material. When a well is subjected to a continuous pumping after a down time, very often there is an initially high TSS as the sloughed biofilms and plug material is pumped from the well. Once this has happened, the water may stabilize with a relatively low TSS value. There are techniques to use this phenomenon to measure the extent of plugging around a biofouled well. They are known as the Biofouling Assessment Quality Control (BAQC) program and involve the controlled pumping of a well in which the plug formation has been stressed to encourage sloughing.

One recently observed event that can be used to differentiate a plugging well from a relatively non-infested (or newly rehabilitated) well is the nature of the bacteria infesting the water column itself. In a plugging situation, it has been frequently noticed that the TSS is not even down the water column but a series of stratified regions that form where different microorganisms have collected and are active. This would mean that a series of TSS readings down such a borehole would show changes as the samples are taken from various points down the strata. In a relatively pristine well, the TSS remains homogenous down the column and stratifications are not recorded. It has been proposed that this may form a relatively simple method to determine if a well is sufficiently plugged to warrant further investigation. If the well borehole is stratified (measured as either the TSS or microbial populations), further investigations would be warranted to establish a rehabilitation/ preventative maintenance program.

2. Mean Particle Size (microns)

Mean particle size provides the mean diameter of all the recordable particles and forms a reference point for the possible origin of the suspended particles. From the experiences to-date, the mean size can be extrapolated to the following possible conclusions:

≤0.5 microns, very small particles may involve some clays, colloids and ultramicrobacteria.

0.5 - 2.0 microns, small particles may contain some bacteria but also clays and colloids may be present or dominate.

2.0 - 4.0 microns, clays and silts may be present but also some bacteria, determine whether there are some particles in the diameter range of 8 to 32 microns which would support the hypothesis of a significant bacterial population.

4.0 to 8.0 microns, suspended particles include a significant bacterial population.

8.0 to 32.0 microns, particles are large and represent sloughing elements from biofilms and/or maturing plug formations. There is a probability of large bacterial populations being present.

>32 microns, massive sloughing and break-up occur in the matured plug formation, possibly with relatively low bacterial numbers.

There may also be significant silting occurring.

In general, as the particle sizes get larger on average, there is a greater likelihood that mature plugs are sloughing into the water. These particles are relatively dense and may settle quickly in static water.

3. Percentage Volume Distribution to the Allocated Sizes

On some occasions, the suspended particles may be found to have relatively specific sizes and a large percentage of the TSS may be seen to have been collected into relatively few allocated sizes. These sizes are referred to as the "bins," each of which will include a relatively narrow range of particle sizes. When bacteria are present in a water sample, they may, in fact, be of relatively similar size and so the percentage volume distribution will be concentrated within those allocated (bin) sizes. Commonly, bacterially dominated particles have similar particle diameters. For example, strains of the *Pseudomonas* species may commonly be found in bins ranging from 8 to 16 microns with one or two of those bins showing the greatest number of particles. Filamentous, stalked and sheathed bacteria tend to rotate in the water. When being detected in the laser beam, the results may show anomalous readings that are often found in the bins from 16 to 32 microns. This is because the laser beam detects the rod-shaped cells and stalks obliquely in the light path and ascribes a diameter which more resembles the length of the rod, sheath or stalk rather than its diameter. When examining a bar graph of the distribution of the volume (or mean cell diameters), it is common to notice a "spiking" where some bins record a number of particles while the bins on either side show none detected. This "castellation" is very typical, particularly for the stalks of the IRB *Gallionella* where these spikes may be seen 3 or 4 microns apart over the range from 16 to 32 microns.

4. Surface Area and Wet Weight *(by predetermined density allocation)*

Most laser-driven particle counters can compute the surface area and wet weight of the TSS. This allows the user to obtain an understanding of the degree of interaction that may be expected between the suspended particles and the water (surface area interface) and also the percentage incumbency of microorganisms within the suspended material through relating viable counts to the wet weight of the suspensions. This latter feature can be useful when appraising the potential involvement of suspended microbe-containing particles in remediation functions (i.e., the greater the density of microorganisms in the particles, the greater the likelihood of biological activity occurring).

5. Temperature

Temperature is a major factor thought to influence the rate of biofouling. Most microorganisms grow within relatively narrow ranges of temperature and this is reflected in the rates of plugging generally experienced in the field. The relationship between plugging and temperature is difficult to universally establish but common guidelines are as follows:

< 4°C	- Very slow plugging by limited range of psychrotrophs
5 - 12°C	- Slow plugging by broad range of psychrotrophs
13 - 18°C	- Plugging by a mixed psychrotrophic/mesotrophic flora
19 - 28°C	- Plugging can be rapid caused by mesotrophic flora
29 - 39°C	- Rapid plugging can occur
40 - 46°C	- Plugging may be rapid with narrow spectrum flora
47 - 75°C	- Plugging generated by narrow thermotrophic flora
>75°C	- Plugging generated by limited range of thermotrophs.

Note that psychrotrophs flourish at <15°C, mesotrophs within the range of 15 and 45°C, while the thermotrophs function within a 20 C° range at temperatures of >45°C. The temperature of the water from the well, therefore, can be used to obtain a general understanding of the likely rate of well plugging.

Changes in water temperature being pumped from a well may also be indicative of a plugging process developing. If the plugging process is changing the origin of water flowing into the well, the temperatures of the product water may change over the time of pumping. If the plugging causes water to be redirected into the well from shallower recharge zones above the well, the temperatures of the product water may reflect the seasonal changes experienced in that water. Where water is redirected by plugging to the well from deeper ground water strata, the temperatures may shift upwards or downwards depending upon the temperatures of the ground waters in those formations.

There is some evidence that microbial activities associated with active biofouling including plugging may also generate heat and cause registerable changes in the water temperature. Research is currently underway to determine the significance of this event to the diagnosis and location of plugging events in and around wells.

6. Pump Rates and Drawdown

Through all of the stages of plugging there are effects on the rate at which water can be produced from a well. Well production capacity is set as the rate at which a well can continuously produce water. This rate of production involves a drawdown in the water level within the borehole to a position where stability is achieved. Very frequently, the demand for a well is only a fraction of the production capacity of the well. The net effect of this reduced demand is to make it more difficult to record the early stages of plugging via the loss in production capacity by a well. For example, if a well had a production of 1,000 liters per minute and was being pumped at a rate of 500 liters per minute, the plugging would have to reduce the flow into the well by 50% (500 liters) before there would be any stress created in the production of water from the well. What is of particular concern would be that a well already 50% plugged in terms of production capacity has a high probability of becoming totally plugged. Unless a well has been operated at close to production capacity, the use of drawdown and the rate of pumping may not be taken as being directly indicative of the rate of plugging.

In the "Armstrong" scenario, a well was pumped to a standard drawdown which controlled the production of the well. A cyclic fluctuation in the production from the well was noted which reflected the triphasic stage of plugging that was occurring around the well at that time. Daily production from this well was noted to shift upwards (facilitated flow) and downwards (restricted flow) by 2 to 5% of the mean flow rates. These shifts in flow were due to the expansion, sloughing and stabilization of the developing plug and functioned over an eleven-day cycle in this case.

Other laboratory and field experiences have revealed similar harmonic occurrences based on cycles lasting between 9 and 120 days with many of the laboratory plugging microcosm experiments functioning in cycles of from 28 to 35 days. This phenomenon may also be observable by monitoring the drawdown to a standard rate of pumping. The drawdown would not be so great during the facilitated flow phase and would be greater than the average during the restricted flow. Such phenomena cannot, however, be easily recorded when the well is not operating at close to production capacity (e.g., >80%).

7. Redox

It has been a common experience in both field and laboratory studies to find the bulk of the microbial biomass forming at the redox front. This is where the oxygen concentrations are very low (i.e., barely oxidative), rather than in the deeper parts of the ground water systems where there is no oxygen and the conditions are often reductive. The nature of the reductive and oxidative status of ground water can be determined as the redox potential that is measured in millivolts. Most aerobic microorganisms function well over the redox range of +50 to +200 and become inhibited over the range from -50 to +50. This zone from -50 to +150 millivolts is sometimes referred to as the "redox front." The redox potential for water sampled from a borehole can give an indication of the likely position of a plugging formation within and/or around a borehole. For example, if the redox potential from the borehole sample is measured, the following possible conclusions could be made (redox in millivolts followed by comment):

> +150	- Plug likely to be deep set in porous media
+50 - +149	- Plug likely to be shallow set in media
-50 - +50	- Plug likely in borehole, any pack and pump
< -50	- Plug likely downstream from well

Unfortunately, in a stratified biofouling within the borehole, the redox front may shift dramatically down the water column from (for example) +150 to -20 millivolts. In such a circumstance, the position from which the sample for measurement was taken will critically influence the data that is obtained. Interpretation of the redox potential, therefore, needs to be carefully considered in the light of the possible interference factors that may effect the value of the data.

8. Direct Visual Inspection

While it may not be practicable to scuba dive down many wells, the use of the down hole camera to log the conditions of the casing and screen within a borehole is becoming a very popular and necessary method for logging a well. There are four major forms in which plugging may be "viewed" down the borehole, but it must be remembered that the only plugging that becomes easily visible is that within the water, growing over the casing and screen and within the slots and immediate porous media behind the screens. Absence of any visible evidence does not mean that no plugging is occurring. The forms of plugging are:

a. Free-floating

Clouded water and the presence of floating particles usually with very ill-defined edges indicate the probability that there is a considerable amount of microbial growth present within the water column itself. Sometimes this growth can become so dense as to black out the light used to illuminate the casing and screen.

b. Mucoid Tubercles

When the plug is forming just behind the well screen and is growing rapidly, mucoid tubercles (resembling in shape "cabbage" or "cauliflower") grow into the well through the screen. These growths are often very fragile and break apart very easily when knocked with the camera housing or lights. The water can become so cloudy that visibility is completely lost. These slimy in-growths into the borehole signify the presence of a plug formation that is probably growing for the most part outside of the borehole in the porous media in close proximity to the well.

c. Hardened Plates

These may be seen as plates of rust hanging down the casing and/or the screens. Unlike the tubercles, the plates "rusticles" are relatively rigid and can resist minor knocks and abrasions that may be caused by the equipment installed or admitted to the borehole. Such plate-like growths indicate that a matured encrustation is present, and the plugging process may be well advanced. These plates represent the inner edges of a matured plug formation that may extend for some considerable distance back into the porous media around the borehole.

d. Covert Plugging

The first impression when there are no signs of growth in or around the borehole, is to assume that there is no plugging taking place. While this may be the case, there remains the possibility that there is a covert form of plugging forming beyond the field of vision. That is to say, it is growing within the porous media surrounding the borehole. Relatively high (oxidative) redox values (i.e., >+ 50 millivolts) for the water, coupled with a knowledge that there are likely to be nutrients reaching the well in sufficient quantities to stimulate biofouling, can be taken as the positive indicators of the probability of plugging occurring within the formation.

F. BIOLOGICAL FACTORS INFLUENCED BY BIOFOULING

It is generally believed that the majority of the biological factors affecting biofouling and plug formation are bacterial in origin. In the deeper fouled zones where the redox front resides and beyond, this would certainly be the case. However, in oxidative environments (redox > +50) there may also be significant populations of protozoa feeding on the bacterial growths. Occasionally, micro-algae may also be present. These micro-plants are normally considered to be plants and obtain their energy from the photosynthesis of light. Many have also been found to be able to survive and even flourish in the dark, using organic sources of carbon. Molds (another major group of micro-organisms) can also be recovered from well waters that have been recharged from unsaturated zones within the aquifer. When this happens, the ground water may carry very high spore populations. Spores may bias the attempts to enumerate the microorganisms present in the water by overgrowing the agar plates. Earthy-musty odors are sometimes associated with the presence of these organisms and, in particular, a group of bacteria called the *Streptomyces*.

Enumeration of the microbial populations within a biofouled and plugging water well is made particularly difficult for a number of reasons. These include:

1. Attached Habitat

Most of the microorganisms present in a biofouled well tend to grow attached to surfaces either as loose slimes or maturing plug formations. The organism in suspension in the water itself tends to aggregate into suspended "colloidal" masses, which may contain a mixture of different strains functioning in a consortial manner. Only a relatively few bacteria will grow ("swim") independently within the water column. This creates a significant problem when attempting to determine the scale of a microbial population within biofouling water wells. It is quite possible for water drawn from a heavily biofouled water well to contain no detectable microorganisms simply because all the organisms are attached and not actually suspended in the water as such. Even if there were microorganisms in the water, the fact that they are commonly in a suspended "biocolloidal" form means that enumeration becomes difficult using the classical spreadplate techniques in which the bacteria are grown up on an agar surface to form separate and distinct countable colonies.

2. Oligotrophic Nutrient Regime

Oligotrophic means "slight" growth due mainly to a shortage of suitable nutrients. Most of the microorganisms are in a state of nutrient starvation and relatively traumatized. When such microorganisms are taken in a water sample, there is often a recovery period which is necessary before the organisms can begin to grow. Such traumatized organisms are often not enumerated correctly, simply because the culturing (incubation) times are kept to a minimum in order to "turn around" the samples in a reasonable time for the users. Better recoveries are achieved where the cultural conditions are changed more slowly thus allowing the organisms time to adapt to the richer and more nurturing nutritional regimes that are created in the laboratory.

3. Erratic Population Recording

There are so many variables in the enumeration of microorganisms in water, ranging from the effects of stratification to randomized and periodic sloughing. These events cause the numbers of microorganisms to be recorded in the water to vary depending upon on their degrees of dispersion between planktonic, biocolloidal and sessile states.

4. Specialized Growth Requirements

Many of the microorganisms within the water may require very specific environmental and cultural conditions for growth that may not necessarily be generated when the laboratory attempts to culture these organisms. Very often total counts of microorganisms in water using techniques such as the Acridine Orange Direct Count (AODC) method reveal far more microorganisms than are enumerated using the classical agar spreadplate techniques. The AODC technique, when compared to the agar spreadplate media methods, commonly generates three orders of magnitude higher counts.

There are a wide variety of the agar media that have been used to enumerate bacteria. Popular agar culture media include R2A (heterotrophic bacteria), Glucose Tryptone Soy (fastidious heterotrophic bacteria), WR (iron related bacteria) and potato dextrose (molds). In general, there has been a lack of consistency in the data generated using the agar spreadplate. The variability in the data so obtained has led to a loss in confidence when attributing a plugging event to a microbial origin. As a result, alternative techniques have been developed in an attempt to improve the reliability of the data.

One simple technique that has recently been developed and patented is the biological activity reaction test (BART^{TM}) which compensates in some manner for the various short-comings in the present techniques applied to determine microbial challenges in ground water. The basic premise is to eliminate shock during the transfer of organisms from their natural environment to the laboratory conditions. This is done using the concept of the Winogradsky column in which redox conditions are established from oxidative (top of the column) to reductive (bottom of the column). This is exaggerated in this test by the insertion of a floating ball to reduce the oxidation zone. Nutrients to support growth gradually diffuse up from dried pellets in the base of the tube. Counter gradients are, therefore, created with nutrients diffusing upwards while the oxygen diffuses downwards. Microorganisms may, therefore, locate and grow at suitable sites along these gradients within the water sample.

Plate One, Severe biofouling can include corrosion usually caused by sulfate reducing bacteria. Commonly, the slime coatings will be blackened and the impacted metals pitted and corroded away.

Plate Two, Pumps and pipes can become encrusted with brown concretions which plug up the passageways and coat surfaces. These growths are predominantly iron-related bacteria.

Plate Three, Laboratory mesocosms were used to refine the technologies applied in the BCHT™ process. This unit consists of four stacked gravel packs each with a 2" central slotted well screen. It is driven by air-lifting growth media to stimulate iron-related bacterial plugging. Plugging can be achieved in 120 days.

Plate Four, Early BCHT™ trials in 1987 and 1988 employed a single mixing tank, a 750,000 btu/hr boiler with a tripod support for the downhole treatment of the well. Grenada Dam relief well trials.

Plate Five, By 1994, the BCHT™ employed a self-leveling truck with a rotating extendable boom towing a flat bed trailer with three mixing tanks and a 1,400,000 btu/hr steam boiler.

Plate Six, The truck was used to tow the flat bed trailer and aid in moving the stocks of chemicals such as the drums of CB4 (foreground), while the rotating boom was used to lift and lower equipment down the well being treated.

Plate Seven, The flat bed trailer was moved periodically back to a central depot to reload the three mixing tanks with water and chemicals. The compressors and steam generator are to the right of the tanks.

Plate Eight, The rotating telescopic boom (shown extended over a well site) allows the treatment tools to be inserted in the borehole conveniently. At the mid-point of the extended boom is an apparatus designed by G. Alford that allows the automatic raising and lowering of the treatment equipment downhole.

Plate Nine, The precise positioning of the treatment tools downhole can be controlled and moved using a rotating wheel with a concentric piston driving a pulley positioning system. By adjusting the rate of rotating wheel and the distance to the concentric pivot, the speed and length of stroke can be adjusted.

Plate Ten, Bi-directional jetting tool head commonly used in the BCHT™ application. The upper (to the immediate left of the larger nut) jets eject hot solutions at a 45° angle upwards at pressures from 250 to 700 psi. The lower jets angle downwards to achieve a similar effect.

Plate Eleven, During the BCHT™ treatment, the color and turbidity of the effluent can be used to determine the effectiveness of the treatment at that time. Clouded brown effluent (above) would indicate that iron-related bacterial plugging is being discharged.

Plate Twelve, As flows increase and the water improves in clarity and is no longer discolored, the discharge (above) indicates the effectiveness of the treatment.

Plate Thirteen, Air-lifting can effectively surge the loosened plug material from the wells. The water will often "fountain" out of the borehole with a brown color, the plug material being high in iron.

Plate Fourteen, As air-lifting continues to surge the debris from the borehole, the water may change to reflect the amount of debris being removed. Once most of the debris has been surged, the water will clear (above) showing that the removal of disrupted material has been completed.

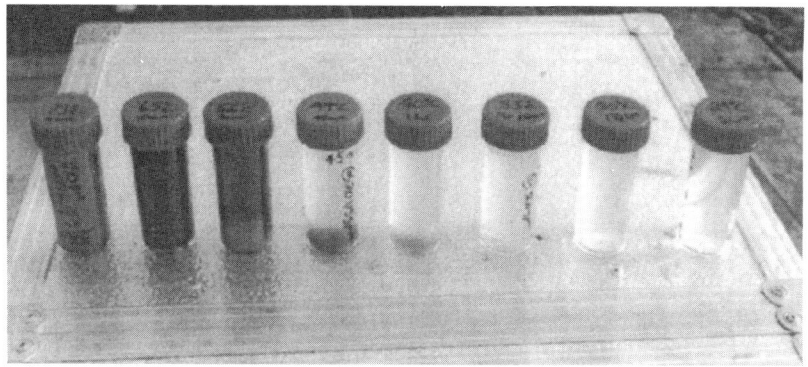

Plate Fifteen, Sampling the discharge water over time (60 ml samples shown from left, earliest to right, latest) can be used to chemically and biologically determine the effectiveness of the treatment.

Plate Sixteen, Watching the discharged effluent from the treated borehole during post-treatment operations can give a good indication of the amount of debris that has been removed from the well as a result of the treatment.

2

APPLICATION OF HEAT AND CHEMICALS IN THE CONTROL OF BIOFOULING IN WELLS

I. APPLICATION OF HEAT, BACKGROUND INFORMATION

To begin, this chapter has been designed around the application of heat, along with chemicals, to achieve rehabilitation water wells that have become plugged. Traditionally, water wells were thought to be plugged with silt, sand, and clays that had often been concreted into various forms of encrustation. Research over the last twenty years has shown that many of these plugging events have actually been driven by hardening biological growths which have then plugged the well. Hence, I will be using the term "plugging" to more appropriately reflect the events occurring in biologically fouled wells that are loosing specific capacity. The U.S. Army's Corps of Engineers have, traditionally, had to rehabilitate such plugged wells. Recovery rates were often poor, since the treatments being applied commonly involved the cold application of chemicals without any real understanding of the root causes of the production problems.

In the 1970s, the role of microorganisms in the losses in specific capacities of wells was hardly acknowledged, let alone understood. Plugging was viewed to be a physical-chemical phenomenon through which materials were moved closer towards the wells and then perched within the throats of the pores in the media around the well. Subsequent chemical activity caused these materials to form into concrete-like structures (encrustations) and, essentially, seal off the well. This process would cause a gradual degeneration in the specific capacity of the wells, followed by a sometimes catastrophic failure (the well had gone "dry").

Microbiologists working with microbes in geological formations have often noted evidence of iron-related bacteria in iron-rich concretions, encrustations, and even in bog iron-ore deposits. For example, David Ellis in the classic work on "Iron Bacteria" published in 1919 made many observations on these linkages:

> "Molisch has found four bog iron-ores which, either in parts of, or on the whole of, the surface, or, in the case of one ore, throughout its substance, contained abundant remains of *Gallionella ferruginea* and *Leptothrix ochracea*, two of the best known iron bacteria.... The formation of iron-incrustations on iron surfaces exposed to water is indubitably influenced by the activities of the iron-bacteria...iron-bacteria are responsible for the formation of ferruginous sediments which probably form themselves by slow degrees into iron-ores...our knowledge is still somewhat meagre.... Enough has been shown, however, to make us beware of assuming that in no case have organisms played any part in the foundation of ferruginous ores."

This book, however, was not required reading for engineers and geologists, and these comments went unrecognized as to their true significance. A major problem that David Ellis reported in the book was the form of iron bacterial infestations as "slimy ferruginous streamers, ferruginous tubercles, ferruginous incrustations of a more or less lumpy nature, and changes of the iron into a mass of a spongy nature".

These types of growths are commonly seen down iron-related bacterially infested water wells. Some impacts reported by David Ellis in 1919 also ring true for the problems being experienced with iron-related bacteria today. These were summarized as:

1. Increasing the amount of organic matter in solution in the water.
2. Decreasing the bore of the conduit pipes.
3. Decreasing the velocity of flow of the water.

It is, perhaps, important to read the opening statement in the preface of this book where David Ellis states: "It is nowadays increasingly felt that there should be greater co-ordination between the work of the biologist and the application of this work by engineers and chemists". Ellis summarizes the problems by stating:

> "There is still a lingering doubt in the mind of the *Practical Man* that science and practice are antagonistic, and in the work of the past, in so far as this touches the iron bacteria, the outcome has not been the best that was possible, because problems have been attacked divorced from biological considerations."

It is now 1998, seventy-nine years later, and only now are the first steps being taken to integrate a knowledge of microbiology into the management of water wells being biologically fouled primarily by iron-related bacteria. From 1919 until the early 1980s, there has been very little research or intensive studies to bring into sharp focus the true relevance of iron-related bacteria to the losses in efficiencies in all types of water wells. This has now started to change with the Regina IPSCO Think Tank in 1986, the American Water Resources Association International Symposium on Biofouled Aquifers in Atlanta, 1986, and the double symposia given in 1990 in Cranfield, U.K. on microbiology in civil engineering and the management of water well maintenance problems. There has not been a major event since that time but there is now a greater recognition of the problems on a global scale.

One thing that is now being realized is that iron-related bacterial biofouling is not limited to just Europe and North America. A survey conducted by Cullimore and McCann in 1979 found that there were iron-related bacterial infestations in all the continents. Particularly severe fouling of wells has been reported in India, the Middle East, Australia, the whole of North America, continental Europe, Africa and China.

Essentially, wherever the waters are iron-rich then it can be expected that there will be iron-related bacterial (IRB) problems. This does not exclude the serious nature of other bacterially generated problems in ground water and water wells. These include sulfate reducing bacteria (SRB, corrosion, taste & odor problems), pseudomonad bacteria (hazardous waste sites, bioremediation, organic pollutant degradation), enteric bacteria (septic seepage, coliform-hygiene risks), fungi (mat formation in semisaturated zones) and the nitrifying bacteria (oxidation of ammoniacal putrefactive products). These problems do not have to be treated as separate events but can occur concurrently. We need the "David Ellis" approach now in order to address the relationships between water well performance and the microbiological factors that can both help and harm this performance.

Over the last decade, a "trickle down" effect has started to gain momentum. Here, the unrecognized research in a few laboratories is now becoming embraced, and we can see this in the practices of water well management even today. For example, the biological activity tests (BART™, Droycon Bioconcepts Inc., Regina, Canada) are now being widely adopted as a simple tool to determine the presence and aggressivity of many of the nuisance bacteria. In Canada, the Canada Agriculture Prairie Farm Rehabilitation Administration-Technical Services (PFRA-TS) are in year three of a seven-year project to determine the extent of water well biofouling. In the U.S., the American Water Works Association Research

Foundation is supporting a very major initiative to determine the most effective ways to rehabilitate water wells. This includes an appraisal of the extent and consequences of microbiological fouling in water wells. In this workshop, the blended use of heat with chemicals will form the main focus. To do this, the workshop will include a historical context followed by a summary of the critical developments in the blended use of heat and chemicals that primarily occurred between 1989 and 1993. After this, a discussion of the present knowledge will follow and the likely pathway developments will take in the foreseeable future. As this "trickle down" continues, there inevitably will have to be recognition of the economic costs of not rehabilitating a water well, compared to the economic advantages of rehabilitation which carries the added bonus of making the well a more sustainable investment.

II. HISTORICAL ASPECTS OF THE APPLICATION OF HEAT

My first direct exposure to the application of heat as an aid in the rehabilitation of biofouled wells was in early 1981, when the topic of the "pasteurization" of plugged water wells came up. The idea, developed by Roy Cullimore in the seventies and first successfully demonstrated in 1975, was to simply heat up the water well environment and kill the infestation off with heat. Well, needless to say, sometimes it did work and sometimes it didn't! The concept was simply an extension of the Louis Pasteur's idea that spoilage (in beer, wines and milk) could be controlled simply by the application of a heat shock. Unfortunately, the types of biofouling in and around a well are more difficult to remove than can be achieved simply by heat.

Through the last three decades, there has been a growing understanding of the manners in which water wells can become biofouled that can affect performance characteristics. There has been a considerable amount of confusion over the terminology used to describe the cause of this degenerating performance (usually, most conveniently, specific capacity is taken as the "benchmark" for performance). Terms commonly used include:

1. Clogging

A chemical-physical blocking of the throats of the porous media leading into the well. This involves an ongoing collection of particles that are too large to pass through the voids into the well itself. Essentially, these materials are simply "perched" around the well.

2. Plugging

A blocking of a well through some form of occlusion which appears to be growing in mass rather than simply an ongoing accumulation of inanimate material. This growth may take the form of slimy, concretious (lumpy), or spongy masses which can form both within the well borehole, on the screens, and back in the formations of porous media beyond.

3. Bioplugging

A blocking of a well in which there are clear signs that the growths involve a biological activity. Commonly, this activity can be caused by bacteria belonging to one or more of the groups: iron-related bacteria (IRB), SRB, slime-forming bacteria (SLYM), heterotrophic aerobic bacteria (HAB), denitrifying bacteria (N), and coliform bacteria (COLI). Often, these bacteria grow in complex consortia within slimy or concretious structures. Where the plugging extends out around the well, this may be referred to as a "bio-plume". It should be remembered each plug is likely to be composed of a number of biofilms that are parts of the growth.

In this work, the term "plugging" will be used since this one defines the loss in the wells specific capacity, as involving a major biological factor defined above by the terms "plugging" or "bio-plugging".

How the biofouling, as such, sets up around wells to cause plugging remains a subject of some speculation, but scientific understanding is developing. The basic approach being used to determine the nature, form and function of plugging is now becoming the subject of more scientific studies. These studies range from:

- Laboratory simulations of water well conditions in what are known as "mesocosms";
- Intensive studies of the dynamics of plugging through sampling water wells which are at some stages in the plugging;
- "Pure" culture studies of the bacteria isolated from, or thought to be involved in, biofilm formation.

This does not preclude the direct investigation of plugging water wells. Wlad Wojek, a Polish engineer, working with the operation of dewatering wells around a strip mine in Krakow observed that these wells were suffering from a plugging caused by IRB. As the strip mine was expanded, the wells were taken out and re-used. This gave Wlad an opportunity to literally "dissect" the wells vertically *in situ* (a rare opportunity!). Carefully slicing down and then cutting them out revealed that the IRB had, indeed, formed complex dark brown ochrous growths around these wells. These growths were either

- Tight growths close to the well screen which do not extend far out away from the well;
- Tight growths close to the well but there are finger-like extensions that may ribbon outwards from the plug for distances of (commonly) 2 to 5 feet. These extensions, generally, are much more pronounced in the direction from which the ground water has been flowing towards the well. This is sometimes known as the "Starfish" or the "Spiculate" forms of growth;
- The maximum growth does not occur right at the well, but is set back as a cylinder of ochrous growth around the well. This set back may be as little as a few inches or as much as several feet away from the well casing. The edging is often spiculate on the outside and relatively smooth on the inside;
- Dispersed brown ochrous deposits were also observed commonly in regions affected by surface water recharge. Here, the higher nutrient loading would influence the ground water and the plugging. Also, the oxidation states in the recharge water could influence the aggressivity of the biofouling. The major result of this impact would be that plugging would become dispersed throughout the zone of influence from the recharge water. This is, incidentally, a major problem for water wells set back from surface waters such as rivers and lakes — a problem from the point of view that the bioplume is more dispersive and, as a result, more difficult to treat.

Experiences of some of the water well managers in Belgium have given an insight into how the ground water enters a well. Generally, when a pump is running up in the casing of the well, the majority of water is coming in from the upper 15 to 30% of the screens. The rest of the water entering the well is moving more "lazily", since there is less demand for this water. Some ground water permeating into the well may actually be moving straight up the gravel pack before entering the well. This has important implications on the form of plugging that occurs around and inside a well.

In the research with the U.S. Army's Corps of Engineers, we were able to take stratified samples down the water column of the well itself. It was also possible to go down through the gravel packs, take stratified samples there and then move out one foot, three feet, nine feet on out for about fifty to one hundred feet around the well. This could be repeated after three months, six months and then in nine months. It now becomes possible to observe the accumulation of organisms, organics and other nutrients, and metals coming toward the well.

III. PASTEURIZATION OF WATER WELLS

In the years from 1971 to 1988, most of my experiences were with pressure acidization in the wells. The results of these treatments were not as successful as would have been hoped, and there was a need to generate some improved treatment technologies. In the course of searching for improvements, the work of Roy Cullimore on the application of heat to rehabilitate water wells came to my attention. Essentially, the concept of heating a well was an extension of Louis Pasteur's concept that spoilage could be controlled by the application of a heat shock to traumatize the spoilage microorganisms. Both in the laboratory and in the field, this application of heat had success. In cooperation with Roy Cullimore, a number of different techniques were employed to heat the water wells. These ranged from: (1) the injection of steam, (2) batch injection of hot water, (3) permanent installation of immersion heaters at the well screens, (4) sequential application of heat (tyndallization), and (5) applying microwave radiation to the wells. Essentially, all the above treatments had successes but also an unacceptable number of failures. Most of the failures arose from a phenomenon in which the heat caused the plug to "coagulate" like the white of an egg. The result of this coagulation was that large sheets of polymers would tear away from the treated area and plug up the throats and the screen slots of the wells. This type of plug was very difficult to rehabilitate. From 1984 to 1989, the treatment went "back to the laboratory". It was clear that heating water wells, while achievable, would not generate a positive outcome (recovery of specific capacity) with enough certainty to make the technology acceptable.

The outcome of these years "in the laboratory" was a certainty that there had to be a blend of strategies which would, when combined, create a synergy in which the final treatment exceeded the expectations of each of the components in the treatment. By 1987, the answer was a blended chemical heat treatment (BCHT™, ARCC Inc., Daytona Beach, Florida). It is common knowledge that one of the effects of heat is to make most chemical reactions go faster. A good example is when washing the dishes; it is a lot easier to get suds with a dishwashing liquid (surfactant) with hot water rather than cold. The concept was to apply that same simple principle to the cleaning of plugged water wells rather than dirty dishes! As the laboratory studies proceeded, it became apparent that aspects of the cleaning (of water wells) had to include heat, a detergent and also some means of making the pH unsuitable for the survival of the microorganisms in the plug.

It has to be remembered that a plug consists of surface growths of slimes coated with, and containing, large concentrations of accumulate commonly including iron, carbonates, phosphates and silicates. All of these elements serve to protect the biofilms that are the "homes" for the different microorganisms living within the growths. To clean all of these different materials off the surfaces is, naturally, much more of a challenge than simply washing grease and debris off plates with hot water and a dishwashing detergent. Biofilms are at the core of the plug and are forming the structures that make the slimes, encrustations, nodules, and crystals that all coat the growths. For the most part, biofilms glue these structures with polymers. What has to happen in rehabilitating water wells is that there has to be a successful de-construction of the plug. In other words, this polymeric "glue" has to be broken down into small enough fragments that it can be easily removed from the well.

Research centered on the sites of bioplug formation around water wells revealed that, commonly, the growths were heaviest at the edge of the gravel pack. This is, in all probability, one of the first sites for biofilm formation. It is here, where the water was coming in that there is a sudden drop off in velocity of the water moving into the well. From the laboratory studies working with simulations of water wells called "mesocosms", it became evident that this sudden change in velocity caused turbulence and led to the biological accumulation of nutrients, iron, and other materials which formed into the bioplug. While the biofouling may begin near to, or on, the screen, the mesocosm studies showed that such growths would also grade off back into the formation. This means that just because bioplugging cannot be observed when camera-logging a water well, it cannot be assumed that there is no bioplugging "out of sight, out of mind" back in the formation.

IV. BLENDING CHEMICALS WITH HEAT TREATMENT

The BCHT™ emerged from the laboratory studies as a phased treatment between 1985 and 1988. Three phases were considered to be essential to clean the plugging from the surfaces and get a recovery in flows approaching, or equal, to the original specific capacities of the newly developed (and relatively unbiofouled) water well. The three phases became first, to shock the microorganisms inhabiting the bioplug; second, to disrupt the structures of the bioplug (deconstruction); and third, to disperse all of the disrupted materials off the surfaces so that the water well can return to original characteristics.

In the shock phase, a combination of disinfectant, detergent and heat is applied with the aid of a jetting tool to maximize impact. The disinfectant may be chlorine-based or, more recently, one of the simpler organic acids. Using a polyelectrolyte detergent increases the effectiveness of the disinfectant. From 1985 to 1988, the most effective detergent for the shock phase was PM-30. Many of the wetting agents (detergents) tested caused the polymeric sheets making up the bioplug to tear away in large sheets to plug downstream passages and pores. PM-30 was found in the mesocosm studies to be the only one tested which did not cause sheet shearing and coupled plugging of the pores. Since 1996, a more effective wetting agent (CB-4, ARCC Inc., FL) has been tested and found to be even more effective at lower concentrations than PM-30. Today, most of the wells being subjected to BCHT™ employ CB-4.

In this shock phase, the wetting agent has, essentially, to penetrate into the biomass, lift it up and bust it up into a palpable product ready for the second (disrupt) phase. Another thing that is important in the shock phase is to reduce the demand on the chemicals during the second phase and place the microorganisms in the state of physiological shock. It is very important to maximize the penetration of the chemicals in all phases of the treatment, in particular, during the shock phase. Heated chemicals are jetted in with the tool being moved up and down the well with the solution running at between 90° and 212°F. Pressure at the nozzle (depending on circumstances) may be set at up to 750 psi. When the tool is going down into the well, the jetting is going into the slots. Here, debris is being cleaned off the slots and the gravel pack is being penetrated.

The objective is also to set up turbulence in the borehole to improve chemical entry into the plugged zones. Not only is the movement linear (out into the formation), but when this hot chemical solution hits the cold water, a convection process forms setting up secondary motion. Everything now begins to move and flow on its own (the well has started to "Rock and Roll"). As the heat reacts with the cold water, so the convection process moves upward and outward with the pinnacle moving out into the formation. This is how deeper penetrations are achieved. An example of this is the Battleford BCHT™ treatment carried out in the fall 1997 on well #15. Here, that pinnacle of penetration moved out from the treated well by 25 feet or more, and there was a 10 C° rise in temperature and a drop in the pH of more than 2 units. The bottom line is that you have to carry the treatment out into the formation. The other important part of the shock phase is to impact the bioplug to essentially soften it, loosen it up, and get everything moving around. If the screen can be cleaned up during the shock phase then that is going to mean less chemical demand during the second (disrupt) phase.

This disruption phase usually follows an overnight contact time and can utilize the same mechanical procedures as the shock phase or an alternate methodology which is a closed venturi. This operates as a modified venturi that easily fits into a well. It is made of a 4 or 6 inch stainless steel casing material plus joint threads. The steam line used with the venturi runs down to, or close to, the bottom of the venturi device. The steam is then jetted out of the bottom of the venturi. This sets up low-pressure differentials in this area where the temperature becomes concentrated along with the applied chemicals. This essentially concentrates the heat within that area. Generally, this venturi tool is anywhere 6 to 10 feet in length. Moving the device slowly up and down the well allows a more carefully managed application of the heat and the blended chemical.

Disruption is partly achieved by radically lowering the pH commonly to less than 2. The minimum is to have the well with a pH of less than 3 during the disruption treatment phase. Again, the objective is to try and apply the heat and chemical not only within the zone, but also get out far enough into the formation as may be economically possible. It is essential to consider that economic factor. Any treatment being applied to rehabilitate a well cannot go on forever, and there comes a point of severely diminishing returns for the effort being expended.

The degrees of temperature rise achieved at the end of a BCHT™ is critical to the success of the treatment. Generally, a typical treatment will involve a 30° to 50°F rise. If a 50°F rise in water temperature is desired, then the borehole water should be at somewhere above ambient by 80°F. This will normally keep the convection going which is moving the treatment chemicals back further into the formation. There have been occasions when, even 24 hrs after treatment, a visual inspection down the borehole will show the treated water to still be slowly stirring itself as the cold water comes back into the well trying to replace the hot water.

Essentially, in the disruption phase, the bioplug has now been severely shocked and the structures are collapsing. As this collapse continues, so the mass begins to fragment into particles which also fragment. A successful BCHT™ must reduce the size of those fragments so that they can float away from the surfaces and be removed from the well.

The final third step is the disperse phase. Here, the objective is to remove the fragmented biomass from the water well so that it can return towards original specifications. Personally, I view this disruption phase as simply "plain good old fashion well development". This is the stage that is often missing with so many treatment technologies! It is obviously not good enough to simply throw a barrel or two of acid down the hole or, perhaps, just jet away a little bit of stuff in there and then stick a pipe back in, pump it off quickly and then take off. Treatment involves maximizing the impact of the treatment and cleaning out the product of that treatment as efficiently as possible. Remember that the treatment would have, minimally, loosened up all material in the well and some of that would now have piled up or floated along the convection pinnacles further out into the formation. Phase three is to disperse the impacted material preferably by taking it out of the well, so that it cannot form the feedstock for the next round of plugging.

Dispersion, therefore, means that there has to be a physical removal of all of the (disrupted, phase two) material which will, if not removed, start blocking up the well again. There has to be quality time spent with the well during this phase as during the original development of the well. So often, rehabilitation simply means that the contractor has blown a hole through this biomass causing the plugging. When this happens, there will be an increase in flow, but in the short order of perhaps 3 to 4 months, that hole collapses down in on itself due to growth, and now there is the well all plugged up again! What is essential during the final stage of treatment (dispersion phase) is that all of the bioplugging material is removed. This is equally true for any other treatment that has impacted on the biomass down the borehole. It does not matter which treatment, all will produce debris which will have to be removed.

BCHT™ has been developed in the last ten years and many lessons were learned during that period. In retrospect, this development occurred in two stages. The first was a proofing out of the technology which took place between 1988 and 1993 inclusive. After 1993, the question now shifted from proving out the technology to one of developing the technology for mass economical application. These will be addressed separately.

Since 1988, research has been going on mainly in the eastern states on the application of the blended chemical heat treatments under the sponsorship of the U.S. Army's Corps of Engineers. The reason for the interest being generated in that agency was the acute concern about the potential failure of earth-fill dams and levees as a direct result of plugging in the relief wells associated with these structures. In 1988, work was concentrated on the Upper Woods River and Levee, Illinois.

V. BCHT™ DEVELOPMENTS

A. UPPER WOODS RIVER, 1988

The Upper Woods River Drainage and District is bounded upstream by Lock and Dam 26 at Alton, Illinois and downstream by the mouth of the Woods River at the Mississippi River. Relief wells are located along the land side toe of the upstream portion of the levee.

Most of the wells were installed in the 1950s. Each well consists of 8" inside diameter wood-stave well screen, with riser pipes, gravel filter, sand backfill, and a concrete upper backfill. The screens are perforated with 3/16-in. vertical slots; the bottom of each screen is closed with wooden plugs. The tops of the wells are protected with corrugated metal guards and fitted with back-flow valves.

Aquifer sediments at the site are 60 to 110 ft thick with the permeability of sand. This permeability (below the floodplain) averaged $1,650 \times 10^{-4}$ cm/sec with a range around that value of -550 and $+1,200 \times 10^{-4}$ cm/sec.

Plugging was suspected to be a problem since earlier work reported that the wells were declining in effectiveness with the rate of decline decreasing over time. However, as a result of the prolonged flow associated with the flood of 1973, the wells had improved again nearly to their original effectiveness. Three years later in 1976 the wells had an overall average of 78% of original specific capacity but then again began to degenerate. Measures were undertaken by the St. Louis District to redevelop the wells, the expectation being that the wells could be returned to at least 80% of the original capacity. After an initial pump test each well was treated with a mixture of trisodium phosphate (TSP, a dispersing agent) and HTH chlorine as a disinfectant. After 18 to 24 hrs of contact time, the wells were surged using a customized surge block. This surging was performed cyclically, and commonly consisted of fifteen complete passages of the surge block up and down the full extent of the well screen. Some experimentation was used to determine this optimal cyclic pattern. As the treatment continued, it became evident that the realistic recovery was to only 70% of the installed specific capacity. To determine the reasons for this disappointing recovery, two wells (44X and 46) were selected for a more detailed appraisal.

Microbiological investigations of these wells revealed a plug formation was active with an aggressive bacterial population. This was determined to include pseudomonad, sulfate reducing and iron-related bacteria. In the sampled product waters, populations were recorded of >10,000, 500 to 1,000 and >10,000 cfu/ml respectively for each of these aggressive bacterial groups. The high relative population of pseudomonads to iron-related bacteria would indicate that the plug formation was at least partially aerobic and associated with a high organic carbon feedstock in the ground water. These organics would have enhanced the activity of the pseudomonad bacteria.

The treatment scenario selected was based on the site recommendation of Roy Leach (Principal Investigator on REMR Work Unit 32313, Waterways Experimental Station). It was decided to subject the two selected (experimental) wells to the patented Blended Chemical Heat Treatment (BCHT^{TM}) system as a demonstration project using the procedures developed by ARCC Inc., Daytona Beach, Florida. A triphasic (Shock, Disrupt, Disperse) scenario that was adopted used the following stages:

- *Shock.* 800-ppm gaseous chlorine batched to three times the well volume with a 24-hour residence time. This was followed by the injection of a heated chlorine solution (>120°F) and sealing the well.

- *Disrupt.* An acrylic polymer (ARCC II) - sulfamic acid - chlorine mixture was next injected into the well at 200°F. The pH of the water fell to between 1 and 2. Injection was performed using a low volume, high velocity nozzle (nozzle pressures were approximately 500 p.s.i.) working with 1.5 well volumes as the base volume for the treatment. After injection, the well was left for 48 hrs. to allow the disruption of the plugging mass associated with the well. A second disruption was performed using the heated polymer-acid-chlorine mixture. A waiting period of 24 hrs. followed to complete the disruption phase.

- *Disperse.* Was accomplished by the traditional surging technique used in previous redevelopment initiatives. Nine surge cycles (135 passes) the length of the well screen was performed at a rate of 2.5 ft/sec. This rate was selected because it minimized the damage to the wooden screens. One foot of infiltrate was produced in each well and removed with a small centrifugal pump and suction hose.

A degree of rehabilitation was achieved since both wells showed substantial improvements towards the original installed capacity. Well 44X had improved from 52 to 69% with the TSP-HTH technique, but showed an additional 14% recovery to 83% by BCHT^{TM}. Well 46 had improved from 52 to 58% with the TSP-HTH technique but showed an additional 12% recovery to 70% by BCHT^{TM}. While the average improvement by the chlorine-TSP treatment was 14% of the original specific capacity for the 34 wells treated, the BCHT^{TM} employed by ARCC Inc. gave additional improvements of 12 and 14%. This was above the assumed upper limits for specific capacity improvements using the other two (non-BCHT™) techniques. Since the wells had been improved beyond the ability of the previous treatments, a reactive preventative maintenance schedule was developed to ensure the efficiency of the well to maintain its specific capacity. It was decided at that time to begin to pay as much attention to retaining the improvements through preventative maintenance as to the recovery of the well itself. Sustainability became an important consideration.

B. BROOKVILLE LAKE, 1990

In 1990, efforts were shifted to Brookville Lake which is located in southeastern Indiana southeast of Indianapolis (65 miles) and northwest of Cincinnati (45 miles). The lake itself is formed by the Brookville Dam which is a 2,900 ft long earth filled embankment stretching across the east fork of the Whitewater River. The dam has a maximum height of 181 ft and is set in a deeply entrenched bedrock (interbedded calcareous clay shales and hard limestone) valley and subsequently refilled with glacial till and outwash sands and gravel. These outwash deposits form a depth of 100 to 165 ft beneath the dam. A foundation under-seepage control system was, therefore, designed and constructed consisting of an upstream impervious blanket along with a system of downstream relief wells.

These relief wells are designed to relieve uplift pressures at the toe of the dam and to collect and control any under-seepage. To achieve this, 32 slotted wood screen wells with a 12" diameter were installed along the downstream toe of the dam and connected into a common drainage gallery. The wells have a 100% penetration of the foundation sands and gravels and all are installed to bedrock depths. The dam was completed in 1973 with water impoundment beginning in 1974. Monitoring of the relief wells began in 1975 with a regular ongoing monitoring program in-place. In 1986, the total accumulated flow measured was at only 30% of the 1975 levels with individual relief wells ranging from a 33% (minimum) to an 83% (maximum) reduction from the original 1975 flows. This concern led to a video-logging of half the wells to determine any physical signs of the problem. The logging revealed that the well screens were encrusted with bacterial forms of growth, a clear signal that plugging was occurring.

Normal rehabilitation methods were discounted due to the age of the wooden wells and the presence of protruding joint nails in the well. It was decided that these wells could be suitable for the application of the BCHT^{TM} treatment by the protocol being developed by ARCC Inc. of Daytona Beach, FL. Since the video-logging evidence had shown that bacterial encrustations were occurring in all of the wells examined, test borings were conducted around selected wells. High bacterial populations were detected 10 to 15 ft away from the well, indicating that the plugging process was probably present also in the gravel pack and surrounding aquifer. Well RW5 was selected as the test well for the initial evaluation. This well was chosen because the discharge had been further reduced to 19% of the original 1975 flow levels. Immediately prior to treatment, the flow fell further to 42 gpm (14% of original flow). The characteristic of the well RW5 were: total depth, 141 ft; top of screen, 24.6 ft; total length of screen, 116.6 ft; screen material, redwood; slot size, 1/2" x 3"; static level, 13.5 ft; flow/no flow, 97 ft; with the riser encrusted.

The treatment scenario for well RW5 began with the discharge pipe in the gallery being sealed using one of the locking caps usually used for the above-ground well tops. A 12" seal was placed in this cap and tightened down until there was no leakage of water. Shock treatment was next applied as heated chlorinated water (final concentration, 800 ppm in 100 gals.) which was delivered through a jetting head. This head was moved up and down the well to more evenly distribute the chemicals within the well. Once this had been accomplished, the materials were airlifted for three hours to improve the mixing and force the material out into the formation through the gravel pack. The next phase, disruption, now followed through heating the water in the well through a closed loop system with a submersible pump set at 20 ft from the surface. The water was also circulated through a steam generator prior to discharge back to the well at 125-ft depth. To disrupt the plugging, 1,000 gals of the (800-ppm) chlorine water was next mixed with 600 lbs of sulfuric acid and 4 gals of the wetting agent and injected down the well using a specially designed recirculating injection pipe. Recirculation of the chemicals was maintained for three hrs followed by the well being left to sit idle for 24 hrs. The acid solution then became neutralized in the formations.

The final stage, dispersion, was now undertaken by airlifting to develop the well, followed by three hours of flushing. The discharge water at this time was changing between brown and black and seemed, possibly, to relate to the part of the formation being flushed.

To observe the degree of success achieved by the rehabilitation, immediately after the completion of flushing, the flow was recorded and found to be 125 gpm. This was equivalent to 43% of the original flow, essentially, an improvement of 29% over the pre-treatment flow. Given this result on the demonstration well, it may be expected that a more complete treatment of all of the relief wells could create a synergistic effect that would bring the flow rates up to 60 to 80% of the original flows.

C. SHAW AIR FORCE BASE, 1992

In 1992, work was concentrated at a site at the Shaw Air Force Base near Columbia, SC, which had been subjected to a JP-4 fuel contamination, which had entered the aquifer and formed into a plume. To contain this plume, recovery wells (RW4 - RW12) were installed as a part of a JP-4 recovery system. These wells were installed in 1991 but there were a series of biofouling problems that were focussed in the water phase pumps and recovery wells. Preliminary researches revealed that the fouling was biological and extended from the screen, back into the surrounding gravel pack, and even out into the formation itself.

Because of the extensive form of this plugging which was largely of a slimy nature, it was thought that the management of these very aggressive biofilms (forming the slimes) was found to be a particularly challenging problem. This was particularly true when attempting the removal of the free product (JP-4) contaminant.

In the initial evaluation of the biofouling, the recovery wells were first placed into a "rest" cycle for several days. This was designed to cause trauma in the biofilms forming the slimes, cause sloughing and allow recovery of samples containing these nuisance organisms. When the production wells were then baled using a see-through baler, the samples were found to consist of a 50:50 (by volume) stratified mixture of JP-4 fuel (upper layer) and water (lower layer). At the interface between the two layers was a thin red layer. Examination of this layer microscopically, and by BART™ tests, revealed that this layer was composed of a dominant consortium of iron-related bacteria. Similar layers were recovered from all of the other wells indicating that these iron-related bacteria were indeed present throughout the site and may be forming focus sites where the various plugging processes were occurring. Clearly, such activities could impair the recovery of the JP-4 fuel from the plume via the wells.

Given that there was a strong body of evidence that the plug formations were, indeed, extensive and had entered the formation itself, it was therefore decided to attempt to control the situation by applying the BCHT™ treatment.

The form of the plug was confirmed by the generalized growths of iron-related bacteria which were universally evident in the formation of the "thin red line" in the baled samples from all of the wells after they had been "rested". This layering would indicate that these bacteria might have positioned themselves to utilize some of the organic fractions of the JP-4 fuel through maintaining a density lower than the water, but higher than the JP-4 fuel. In the laboratories of Droycon Bioconcepts Inc., Regina, such density adjustment in biocolloids is a common experience in biofouled waters. When these events were observed, either visually or with the laser particle counter, they became referred to as "bio-submarines" and showed a remarkable ability to maintain position within a water column.

Slime formations were also commonly found on the pumps and drop pipes. The pump intakes (particularly for wells RW8, 9, and 12) were also very heavily fouled with either a clear or white jelly-like mass while the drop pipes contained very turbid standing waters.

Due to the history of ongoing problems with fouling experienced by the contractor, it was decided to attempt to control the biological plugging by the application of BCHT™. The standard triphasic format of shock, disrupt and disperse was used.

- *Shock* treatment was applied using two alternative chemical mixtures. RW10 to RW12 received 1,000-ppm acid chlorine, along with sulfamic acid and wetting agent. RW4 to RW9 received acetic acid (as the alternate disinfectant for the shock phase). Each well was treated with 700 gals of a 50:50 water/chemical phase mix. The shock phase chemicals were jetted into the well from the top to the bottom of the screen as a slow up/down feed at <1 ft/sec. For both solution formats (chlorine or acetic), the terminal pH was set at between 1.2 and 1.6. The feed rate was 10 gpm through a 200°F nozzle. After 30 minutes, the average well temperature was 170±5°F with the pH consistently at less than 2.0.

- *Disruption* was performed by the setting of a circulation tube 2 ft above the bottom of the well and was moved periodically at the operator's discretion by units of 5 ft up to the static water level. This was done in order to continue to distribute the heat and residual chemical feedstock over the full height of the water column. Average temperatures were raised to 180±5°F with the pH < 2.0. The next day, the wells were surged at 3 ft/sec. using a 5" open block for 30 minutes. This was done to agitate any residual acid solution and also loosen any biomass that may remain attached to the surfaces. When the wells were sounded the following day, they were all found to have filled in with 3 to 5 ft of a mixture of deposited materials in the bottom of the wells. Samples of these deposits were taken and found to consist of sand and silty red clays along with a small amount of gravel pack. This material was removed during the application of the disperse phase.

- *Dispersion* of the fouling was by air surging. Water was lifted to the top of the well and then dropped into the bubbles created by the air pressure. All of the wells were alternately lifted and pumped for periods of 2 to 3 hrs and then left to pump overnight.

Exceptions to this practice were RW6 and RW7 because these wells continued to produce silt and sand even after the dispersal phase was finished.

An acceptable degree of rehabilitation was achieved with good results in all of the wells, with RW5 and RW12 recovering to >15 gpm, while all of the others returned to >10 gpm. All of the wells went back on-line and were able to continuously pump at the manufacturers specified flow rates (6 - 7 gpm). Fuel recoveries equaled or exceeded the previous recovery rates that had been experienced. One well (RW12) has a larger installed pump and never recovered much JP-4 historically. This well also continues to pump at >11 gpm, but without fuel recovery suggesting that this well is now outside the JP-4 plume.

While these wells were shown to be plugged by IRB biofouling activities, the BCHT^{TM} treatment was demonstrated to be able to effectively control the problem. There was concern that the wells may have been improperly developed (during construction); this would then have reduced well efficiencies. There may also have been channeling inadvertently created and high velocity zones within the formation and gravel pack that would have compromised efficiency and induced biofouling.

These severe biofouling stresses created in the recovery wells handling the JP-4 plume recovery at Shaw A.F.B. reinforce the urgent need to develop a reactive preventative maintenance schedule to keep the wells operating in a sustainable manner. Such a maintenance program is needed to control the future (and inevitable) plugging events that will arise while the JP-4 plume and/or any products remains active. This would assure an extension of the useful period of operation.

It should be understood that if the ground water is rich in nutrients and microorganisms, then it can be expected that the porous media around each well (e.g., gravel pack) can act as a bio-filter. Such "bio-filters" will entrap various chemicals and organics and allow extensive growth (biofouling) to occur. Frequent "backwashing" as a part of the preventative maintenance should aid in controlling bioplugging. This will then continue to allow the "free" passage of the JP-4 and water into the well for recovery.

The following is the chemical protocol for the preventative maintenance (PM) recommended for application to the RW wells at

Shaw A.F.B. (for application to the wells):

50 gal.	84%	Acetic acid
50 lb.	solid	Citric acid
5 gal.	prop.	Poly-electrolyte wetting agent (ARCCsperse PM-30)
300 gal.	potable	Water

For the pump cleaning, which has also been found to be equally important, the following solution was recommended (for application to the pumps):

5 gal.	84%	Acetic acid
5 lb.	solid	Citric acid
0.5 gal.	prop.	Poly-electrolyte wetting agent (ARCCsperse PM-30)
30 gal.	potable	Water

The recommended scenario for conducting the PM is briefly outlined below:

1. Batch chemicals.
2. Remove pump.
3. Invert pump in 55 gal drum containing PS chemical mix
4. Set airline 1 ft from base of well, secure top.
5. Crack air valve to mix WS chemicals.
6. Set WS chemical feed to just below static level.
7. Slowly add WS with continued air feed.
8. Stop air feed when all of the WS chemical mix is in well.
9. Open up air-line to base of well (water rises up).
10. Sequence agitation with every 10-ft shift up the well.
11. Duplicate sequence 10 at each shift
12. Leave well overnight.
13. Surge well at 10-ft intervals for five min.
14. Remove and clean pump (see 3) and then reinstall.
15. Check pH of water, stabilize with lime if required

Pump treatment (2, 3 and 14) should be monthly, or more frequently if low flows and/or high drawdown indicate that there is a need. Well treatment (1-15) should be quarterly initially with the time period adjusted to need.

D. GARRISON DAM, 1992

In 1992, activity shifted to the Garrison Dam. This work was conducted on the West Terrace Gravel relief wells at Garrison Dam, North Dakota. The wells included in this treatment were all wells which had been previously subjected to a BCHT^{TM} treatment by the methods developed by ARCC Inc. of Daytona Beach, FL. Specifically, the wells included relief wells # 58, 59, 60, 61, 62, 63, 64 and 65. This initiative was generated by a concern that bioplugging might again be occurring in these wells in a manner that could lead to a compromise of the whole system.

It was, therefore, decided to conduct a survey of the likelihood of regrowth and plugging using three critical broad criteria. These were:
1) quantification testing of the wells to determine performance;
2) downhole video plugging diagnosis to determine signals of in-hole plugging;
3) BART^{TM} evaluation for the plugging potential through the determination of the aggressivity of the microorganisms within the wells.

After the previous BCHT^{TM} treatment of the targeted wells in September 1990, most of the wells revealed a degenerating flow. Over the next twenty-six months, the changes in flow rates were (well # given followed by the average gpm loss / month): 58, -0.04; 59, +0.21; 60, -0.02; 61, -0.01; 62, -0.05; 63, -0.02; 64, -0.05; 65, -0.16. This was thought to be due to a recurrence if a plugging and down-hole video-logging was undertaken between May 1989 and December 1992. The first survey preceded the BCHT^{TM} treatment, with a second survey being undertaken eight months after treatment (May 7, 1991), while the final survey was taken after the first preventative maintenance had been applied. Brief summaries of the first two surveys are given in Table Two to illustrate the reinfestation and plug formation in the wells after treatment:

BART^{TM} tests were performed two months before the second (post-BCHT^{TM}) video-logging of the wells in May 1991. Positive bio-detection was achieved for the iron-related, sulfate reducing and slime-forming bacteria (IRB, SRB and SLYM, respectively). All three bacterial groups were detected in all of the wells and were found to be at least moderately aggressive (time lag commonly of <4 days for all tests). This was taken as confirmation that the plugging (evidenced by the gradual diminution in flow) involved a complex consortium of bacteria that included all of these groups of bacteria. Because of the diagnostic evidence that plugging had, indeed, returned to impact on the performance of the test relief wells, the ARCC formatted a preventative maintenance program that was instituted for action on the wells from November 2-6, 1992. Pam Madsen and Joe Hughes of GEMRO-ED-GE conducted the PM program. Both operators had completed the OSHA forty-hour safety-training program for hazardous and toxic materials. This was appropriate because some of the chemicals used in the PM were corrosive and, therefore, the handling and storage precautions listed on the manufacturers "Material Safety Data Sheets" (MSDS) were followed.

The pre-BCHT™ depth was frequently so plugged (*) that the video camera could not be usefully extended to the base of the relief well (RW). The depth is therefore given as the effective "logable" (viewable) depth. Depths over which bacterial growth could be observed are recorded in feet over the range where activity was recorded.

Table Two
Illustrating the Impact of BCHT™ on the Plugging of Relief Wells, Garrison Dam

RW#	pre-BCHT™			post-BCHT™ (8 months)		
	Depth	Position	Plug	Depth	Position	Plug
57	56	53-55	low	57+	none	soft dep.
58	37	18-33	minor	38	none	none
59	39	16-28	minor	40	9-15	minor
60	41	21-31	heavy	42	0-17	slime
61	22	13-16	heavy	45	22-31	heavy
62	13*	11-13	plugged	48	0-28	moderate
63	29*	19-29	plugged	48	0-38	moderate
64	21*	19-21	plugged	51	0-51	moderate
65	42	17-24	heavy	54	15-23	minor

Note: RW# (relief well number); left three columns are the data pre-BCHT™ while the three right hand columns are post-BCHT™; depth is given in feet accessible to the video-logging; position refers to the depths at which bioplugging was observed; and bioplug refers to the observed intensity of growth seen using the video-logging.

Plugged growths prevented the useful passage of the video-camera when heavy growths were extremely extensive but did not "fog" the view excessively, moderate growths were often gradated down the profile, and minor growths were limited to relatively small observable patches of growth on the walls of the well. Soft deposits were reported dominating in one well, while another well had copious slimes formed on the walls. Only one well did not reveal any bacterial activity by this technique.

The chemicals selected for this PM program were sulfamic acid, acetic acid, and a polymer detergent (ARCCsperse PM-30). Acetic and sulfamic acids were selected because of their compatibility and relatively low risks when handled appropriately by properly attired personnel. The PM-30 was selected because of its stability in acidic conditions and total degradation that subsequently occurs without creating a major potential nutrient feed-stock for subsequent plug-forming bacteria. The acid blend recommended for the PM treatment (per well) was 30 lbs of 99% sulfamic acid, 15 gals of glacial acetic acid, 3 gals of ARCCsperse PM-30 mixed with 82 gals of clean potable water to yield 100 gals of saturated solution. Chemical batching was done outside in quantities sufficient to treat four wells (i.e., 400 gals in an 800-gallon tank). The pH of the final solution was 1.0. After surging the chemical solution for twenty minutes with a 125 psi (maximum) air compressor, the tank was transported to the wells.

BCHT™ well treatment followed the sequence specified below:
1) RW outfall pipes were sealed with inflated foursquare balls to contain the treatment solutions within the well. pH readings were monitored before and after the PM. This was done to ensure that the wells returned to the normal (acceptable) pH range and that the acidic solutions had been removed or neutralized.
2) The wells were surged and airlift pumped from the bottom to the top to remove any of the more fragile (dispersible) biomass. This was done to reduce the chemical demand that would have been created if this biomass had not been removed from the well. Surging of the well continued until the water appeared to be clear.
3) For each well, the acid solution was applied at a rate equivalent to three times the well volume. Agitation was accomplished using an air-line installed to the bottom of the well. Sufficient air was used to maintain a maximized agitation without air-lifting the water out of the well. Air surging was also used to facilitate the infiltration of the acidic solution into the gravel pack and sand strata while breaking up any plug formations.
4) The treatment solution was left in the well overnight in order to extend the period of disruptive chemical activity to any surviving plug formations. The following morning much of the acidic solution that had been applied was now neutralized and/or diluted to the point that the pH of the well water had returned to between 1.0 and 0.5 of the ambient pH for that water.
5) In the final stage, the objective was to remove as much of any of the remaining bioplug debris as possible. To do this, the

wells were air surged and airlift pumped until the water constantly flowed clear.

6) Once the treatment was completed, the temporary seals were removed and the wells were allowed to flow under normal operational conditions. Flow readings were taken to determine the initial effects of the treatment and a further video-logging was undertaken in December 1992, one month after the treatment.

The degree of rehabilitation achieved was assessed primarily by the video-logging of the wells immediately after the completion of the PM showed a radical improvement in the wells. The immediate effect of the treatment (Table Three) was that all evidence of plug formations within the relief wells was gone. All of the wells appeared to be clean and a typical post-treatment appraisal was: top riser pipe - clean; top of water surface & outfall pipe - clean; top of slotted screen - clean; typical view of screen - clean. The treatment had successfully removed moderate and even heavy plug formations, slimes as well as the soft bottom deposits and minor intrusive growths. In all probability, the wells after the BCHT^{TM} treatment would have rendered similar results but the video-logging was delayed for eight months during which time some plug formations were reccurring.

Table Three
Illustrating the Impact of BCHT™ on the Plugging of Relief Wells, Garrison Dam

RW#	post-BCHT^{TM} (8 months)			post-BCHT^{TM} after PM treatment		
	Depth	Position	Bioplug	Depth	Position	Bioplug
57	57+	none	soft dep.	58	38	none
58	38	none	soft dep.	38	none	clean
59	40	9-15	minor	40	none	clean
60	42	0-17	slime	42	none	clean
61	45	22-31	heavy	45	none	clean
62	48	0-28	moderate	48	none	clean
63	48	0-38	moderate	48	none	clean
64	51	0-51	moderate	51	none	clean
65	54	15-23	minor	54	none	clean

A comparison of the relative effectiveness of BCHT™ and the PM treatment may be made by comparing the rate at which there is a post-treatment loss in flow occurred due to a regeneration of the bioplugging. By comparing the loss in flow as gpm/month over the historical post-treatment periods, the following comparisons can be made (Table Four).

Table Four
Rate of flow change over time after BCHT™ and after a subsequent PM treatment

Well #	Flow Shift (gpm/month)	
	Post - BCHT™	Post - PM
58	-0.04	-0.08
59	+0.21	-0.63
60	-0.02	-0.55
61	-0.01	-0.95
62	-0.05	-0.23
63	-0.02	+0.18
64	-0.05	-0.35
65	-0.16	-1.35

The mean loss in flow over time after BCHT™ was -0.017 gpm/month while for the PM, the loss averaged -0.49 gpm/month indicating that the PM had not had such a dramatic effect in retarding the rate of re-establishment of bioplugging within the well.

There were, in spite of these faster bioplug regrowths, overall improvements in the flow from the wells. These percentile improvements after the PM treatment were (well #, percentile change in flow rates): 58, 11%; 59, 14%; 60, 5%; 61, 7%; 62, 32%; 63, -9%; 64, 40%; and 65, 0.017%. Well #65 gave the most dramatic improvement of over 1,000% over the pre-PM flow. Other wells, however, showed an average improvement in flow of 14%.

E. LEESVILLE, 1992

Also in 1992, BCHT™ was demonstrated at a study in Leesville, Ohio where the U.S. Army's Corps of Engineers owns more than 5,000 relief wells with related drain systems (piping, outlet valves, and regular sand drains beneath soil and concrete structures). A Repair, Evaluation, Maintenance, and Rehabilitation (REMR) program has been underway to evaluate the most suitable rehabilitation and maintenance programs for the ongoing functional security of these wells. It was, therefore, felt desirable to compare the traditional cleaning with new methods that may appear promising. The methods did not include all of the possible options, but were formed from a mixture of economical standard chemical and/or mechanical procedures.

Twelve relief wells were assigned to this study. These wells are used to control the underseepage from the Leesville Dam located in Carroll County, Ohio in the Orange Township. Leesville Dam is an earthen embankment constructed in 1935 to 1937 as a control structure on the McGuire Creek, and these structures created Leesville Lake. Consequent to the construction of the dam, underseepage became a severe problem on the downstream toe of the dam creating wet surface (swampy) conditions. To alleviate these concerns, a 4- to 8- ft thick filter blanket composed of coarse granular material enveloping a network of drainage tiles was installed below the dam site in 1975. In addition, twelve relief wells were set at 75-ft horizontal spacings to lessen the hydrostatic pressure imposed on, and under, the dam. These relief wells were drilled through the filter blanket and the rock toe of the dam into the underlying alluvial deposits. Each well terminated 4 to 7 ft below the surface of the filter blanket and was individually housed within 48-in diameter corrugated metal casings with lock and bolted hinged lids. These wells were designed as uncapped flowing wells in which the overflow from each well drains, by gravity, through subsurface drains in the filter blanket to discharge into a collector ditch downstream of the toe of the dam.

Inspection of the relief wells in 1986 was conducted to determine whether the loss in performance in these wells could be attributed to iron bacterial biofouling (bioplugging) of the wells. Inspection revealed that the tops of the well casings and the floors were covered with amounts of mucilaginous reddish-brown deposits that are typical iron-related bacterial infestations. Unfortunately, there was no historical pumping test data on these wells, so the reduction in hydraulic performance for these wells could not be ascertained. However, there was clear evidence of bioplugging. Wells 1, 2, 3, 11 and 12 did not flow at all, while the remaining wells in the series only flowed when the static head of water behind the dam was at an elevation of 963 ft or higher. Also of note is that the risers for the non-flowing wells were all between 0.3 to 1.0 ft higher than the highest riser for the flowing wells (4 - 10). Wells 1, 2, 3 and 12 were found to be only able to sustain very low pump rates, so are also considered to be at the outer fringes of the water-bearing aquifer.

The form of the bioplug was thought to be dominated by iron-related bacteria (IRB), so the presence of IRB was determined by a number of observation techniques that included:
(1) Direct low power microscopic examination; and
(2) In-well biofilm collectors.

Emphasis was placed on the microscopic identification of the IRB in both techniques, which tended to favor the more easily recognizable filamentous and stalked forms of IRB. Much of the data gathered, therefore, reflects the fouling which may be associable with these IRB specifically. Filamentous IRB were identified as belonging to the *Leptothrix - Sphearotilus* group. Using the in-well collector, the following IRB were observed (see Table Five).

Table Five
Forms of IRB detectable by direct observation techniques in the Leesville Wells

Well #	Depth		
	upper	middle	bottom
1	L	L	+
2	L	L	-
3	L	L	±
4	-	-	-
5	+	+	L
6	+	+	±
7	L	L	L
8	L	L	L
9	L,S	L,S	L,S
10	L	L	L,S
11	L	L	L
12	L	-	-

where L is *Leptothrix*, S is *Sphearotilus*, + represents IRB observable but not identifiable to genus, ± few IRB present; - no IRB observed on the collectors.

These groupings relate to some extent to the geologic materials associated around the specific well sites. For example, groups i and iii are predominantly in a silty to clayey sand, while group ii are found in more heterogenous formations. There was clear evidence that filamentous IRB were infesting the wells, but the lack of comparative production capacity data limited the ability to determine the extent of any plugging formation within the wells. From this data, the wells could be categorized into four groups (wells with this type of IRB is shown in brackets):

1) Dominant *Leptothrix* throughout (# 7, 8, 11 & 12)
2) *Leptothrix* dominant upper strata (# 1, 2 & 3)
3) *Leptothrix - Sphearotilus* mixed growth (# 9 & 10)
4) Non specific IRB present (# 5 & 6)
5) No IRB detected (# 4).

The treatment scenario for this included:
1) a long-chained linear polyphosphate;
2) hydroxyacetic acid; and
3) sodium hypochlorite.

This investigation was based on the initial characterization and evaluation of the site. A three-step chemical treatment program for the relief wells was developed. The selection of the chemicals was based on particular characteristics. For example, the long-chained linear polyphosphate was used as the first step because of two factors. First, phosphates are sequestering agents capable of forming soluble complexes with iron. Second, phosphates have a detergent capability that can disperse mucilaginous deposits, such as those resulting from the growth of IRB. This dispersion can leave any residual bacteria exposed and more vulnerable to the disinfectant action. Hydroxyacetic acid was included as the second stage for three reasons. First, as an organic acid the hydroxyacetic acid acts as a systemic biocide against some of the IRB. Second, hydroxyacetic acid has a moderate capability to chemically dissolve iron rich encrustations that sometimes form within and/or around microbial biofilms. Third, this acid acts as a chelating agent that can solubilize iron through the formation of soluble complexes. Sodium hypochlorite is well known as an effective disinfectant agent against a diverse range of microorganisms and would, as the third step, significantly reduce the surviving population of IRB and other bacteria within the area impacted by the treatment. In the designing of the treatment schedule it was recognized that there may a risk to the structural integrity of the fiberglass casings used in the relief wells and so this step in the proposed treatment was eliminated.

Step One. A commercially available long-chained linear polyphosphate was dispensed into an empty 55-gal drum and dispensed using a small electrically driven metering pump. The volume of phosphate solution used was predetermined at 3% of the well volume. In this case, the well volume was calculated as being equivalent to the water volume within the casing and the filter pack (40% porosity) and then empirically multiplied by a factor of 1.5 to give the total volume to be subjected to treatment. The pre-measured volume of phosphate solution was pumped into the well through a 1.5 in flexible hose. This hose was initially positioned near the bottom of the well, but was raised gradually as the solution was dispensed to mixing. Surging commenced immediately after addition of the phosphate solution using a mechanical surge block. Surging started at the top of the well intake and continued down to the bottom at a rate of 15 ft/hr. The length of stroke of the surge block was approximately 36 in and the rate of fall of the surge block was at approximately 3 ft/sec. This surging with the phosphate solution was continued for 2 to 4 hrs. During surging it was found necessary to bail the well periodically to remove accumulating sediment.

Commercial grade sodium hypochlorite (12% available chlorine) was dispensed to the well water volume at a rate equivalent to a total final concentration of 50 mg/l, by weight, as chlorine. This addition was designed to suppress any immediate regrowth of bacteria stimulated by the long-chained linear polyphosphates (phosphate is a major nutrient commonly found to be in limited supply within wells). Once the hypochlorite solution had been added, the well was surged for two hours using the same technique applied as already described above. Afterward, the well was left overnight and surged again for 30 minutes before the chemicals were pumped from the well.

Step Two. The polyphosphate treatment was repeated exactly as in step one of the treatment, but after surging, sodium hypochlorite solution was now added to obtain a higher dosage of 1,000 mg/l of chlorine, by weight, concentration in the well water volume. Surging now proceeded again for four hours before the well was left overnight. Chlorine residuals were measured (Hach model CN-DT chlorine test kit) periodically to determine the effectiveness of the dosage used. The following morning the well was surged for 30 minutes before the chemicals were pumped from the well.

After treatment, the degree of rehabilitation achieved by the treatment was assessed. However, some difficulties were experienced since the act of surging was accompanied by the releases of relatively large amounts of sand ("sanding"). Average rates of sand produced (indicated by soundings) varied from 0.02 ft/hr (RW# 1) to 0.75 ft/hr (RW# 8). A rate of 0.25 ft/hr translates into approximately 8 pints/hr. This is generally considered unacceptable particularly when a well generates in excess of 2 pints/hr of sand during surging operations. This excessive "sanding" may be attributable to the possibility that the wells were not properly developed at the time of

installation. It could be assumed that the mechanical effects of the surging would have become a significant factor contributing to the increases in specific capacity in the wells (see Table Six). Specific capacities (gpm/ft at a drawdown of 5 ft, interpolated from the data) were found to improve in eight out of the twelve wells. While the mean percentage increased in the specific capacity of the eight wells (RW# 4 to 11 inclusive) which showed recoveries were 236%. There are two restrictors which make it difficult to attribute these improvements to the two-step chemical treatment of the wells. These are:
1) Mechanical surging may have physically dislodged some impedances to flow (i.e., the sand); and
2) Original capacities of the wells were not measured and so there is no satisfactory scale upon which to judge the improvement toward the original specific capacities of the newly developed wells.

Table Six
Specific Capacity in the Leesville wells after chemical treatment

Well #	Specific Capacity (gpm/ft) Before	Specific Capacity (gpm/ft) After	Increase %
1	<10	<10	-
2	<10	<10	-
3	<10	<10	-
4	15	45	200%
5	75	208	177%
6	155	210	35%
7	27	220	714%
8	120	162	35%
9	40	140	250%
10	45	120	166%
11	15	62	313%
12	<10	<10	-

There were some changes in the chemical characteristics of the well waters before and after the two step treatment of the wells. Most marked of these was the radical increase in the dissolved iron and shift in the Eh (millivolts) redox potentials. For most wells, there was a considerable increase in the dissolved iron (Table Seven) which may have been a reflection of the state of the biofouling microorganisms within the water in and around the wells. Before the treatment, much of the iron in the ground water arriving within the biological interfaces surrounding the wells would have been accumulated into the bioplug formations. As a result the product water drawn from the well itself would contain very low levels of dissolved iron. After the treatment, the plug formations (bio plumes) closest to the well itself would have been, minimally, severely traumatized by the various physical and chemical "shocks" delivered by the two step treatment. Under these circumstances, the effective bioaccumulation would be impaired. At the same time, the action of the long-chain linear polyphosphate would be, in part, to solubilize the encrusted and bioaccumulated iron.

Table Seven
Changes in Iron Concentrations and Redox Before and After Treatment In Wells at the Leesville Site

Well #	Dissolved Iron (mg/l)		Redox Eh (millivolts)	
	before	after	before	after
1	0.03	0.12	151	147
2	1.04	0.26	-10	-36
3	0.86	10.16	-126	-60
4	0.05	0.10	30	86
5	0.05	0.51	42	18
6	0.68	49.60	23	21
7	0.29	10.70	45	8
8	0.84	31.30	-52	-58
9	0.78	38.10	-45	-48
10	0.74	9.60	13	-57
11	3.22	35.90	-93	-43
12	3.29	32.40	-132	63

Very considerable increases in the dissolved iron concentration were determined for Wells 3, 6, 7, 8, 9, 11, and 12. No clear relationships could be drawn between the microorganisms recorded in the wells and these very considerable increases in the iron concentrations. The redox (Eh) also varied in the response to the treatment, but again there was no logical comparison that could be made between these shifts in redox and the microbial populations.

All of the wells were tested using the in-well collectors placed in the wells after the two-step treatment. One major shift in the population was the occurrence in some wells of the stalked IRB *Gallionella*. The data obtained from this study is listed in Table Eight. Five wells (1, 4, 6, 7 and 12) did not exhibit any presence of IRB using the in-well collectors whereas only Well 4 did not show any IRB in the pretreatment survey. In the remaining wells *Gallionella* was the most commonly recognized occurring in 12 of the 14 positive detections for IRB while this genus of IRB was not recovered from any of the pretreatment samples. *Leptothrix* was identified in 8 of the samples, whereas it was the most common IRB recognized in the pretreated samples. *Sphearotilus* was identified only in Well 11 from the middle sample, while in the pretreatment it had been very common in Well 10 and present in the bottom of Well 11.

Table Eight
Post-Treatment Observations on the Presence of IRB using the In-Well Collectors in the Leesville Wells

Well #	Depth (ft)		
	upper	middle	bottom
1	-	-	-
2	G,L	G,L	-
3	G,L	L	-
4	-	-	-
5	G	-	-
6	-	-	-
7	-	-	-
8	L	-	G
9	G,L	G,L	G
10	G	-	-
11	G,L	G,S	G
12	-	-	-

Note that the codes used in this table are: L is *Leptothrix*, S is *Sphearotilus*, G is *Gallionella*; and – indicates that no IRB were observed on the collectors.

From the data, it becomes evident that the dominant IRB recovered using the in-well collectors after treatment was *Gallionella*. This IRB was not detected in the pretreatment samples. There are two factors that might explain this event. First, the *Gallionella* would have been recognized by the ribbon-like stalks that may, or may not, have had viable cells still present at the end of the stalk. Such stalks may have been random "floaters" left over from the treatment, which had preferentially attached to the in-well collectors. Second, the stalks may have been detached during the treatment processes from the deeper set bioplug formations further away from the well. This second explanation would also be supported by the lack of recovery of *Gallionella* before the treatment was applied.

The chemical treatment program had an immediate beneficial effect on the hydraulic performance of the relief wells. For those wells having pretreatment specific capacities of >10 gpm/ft, there were very significant gains in the specific capacities averaging 236% above the pretreatment levels. Unfortunately, the lack of data on the original specific capacities for these wells does not allow the true recovery rate to be ascertained. The "sanding" which was experienced during the surging activities may also have had a major impact on the improvements in the specific capacities but it is not possible to attribute the likely impact level.

From the data gathered it would appear that these chemical treatments did not "sterilize" the wells, but rather acted in the manner of a disinfection from which some IRB were recovering to reinfest the well. To prevent a significant microbiological challenge to the ongoing production capacity of the wells, a regular treatment process should be performed as a part of a preventative maintenance program.

No specific PM recommendations were suggested for these wells, but an on-going schedule of chemical treatments and monitoring for the critical chemical and biological parameters was suggested. Concern was expressed that the long-chain linear polyphosphates may pose an additional nutrient source, causing faster rates of microbial regrowths if the phosphate was not effectively removed from the well after treatment.

F. BOFERS SITE, 1993

In 1993, attention shifted to the Bofers site. Through the REMR work unit, there was a further evaluation of using the BCHT^{TM} as an effective control method. Under most circumstances, this patented technique was found during the various pilot projects to be able to achieve an order of magnitude greater improvement than comparable standard or traditional methods. The time span before repeat treatments were needed was also found to become greatly extended after a BCHT^{TM} treatment had been applied. Because of these observations, an interest developed in the use of the rehabilitation technique on severely compromised (biofouled) "plume capture" wells at active superfund sites. The Bofers-Nobel (BN) site was selected for the demonstration project.

This 85-acre site is near Muskegon, MI, and gently slopes north to south along Big Black Creek near the southeast border. The site is a sandy alluvium for about 100 ft underlaid by clayey glacial till (an aquitard) for about 140 ft, then Marshall Sandstone for about 160 ft, and then Coldwater Shale below 400 ft The surface stratum promotes direct recharge. The sand aquifer is of importance as a source for private and municipal drinking water, while the Big Black Creek is an important recreational area. For 33 years, the BN site was used intermittently for industrial chemical production and in 1989 the site was placed on the superfund list while still being operated by Bofers-Nobel Company. While the reporting had been sketchy leading to uncertainty concerning the nature of discharges from the ten unlined lagoons on the site, there was a very serious contamination of the saturated zone of the sand aquifer. This became a direct threat to the local society and the natural environment that could be compounded if the leachate reached Big Black Creek and spread to downstream locations. Hazardous compounds identified within the lagoons and impacted areas included methylene chloride, 3,3 dichlorbenzidene, analine, azobenzene, benzene, and benzidine. All are highly toxigenic; and the first four are potential carcinogens in humans, while the last two are confirmed human carcinogens. About 500,000 cu yd of contaminated soil is involved at the BN site.

In 1976, a 12 purge-well net was constructed to contain the contamination plume on-site. Early records reported losses of efficiency and "encrustations", indicating the onset of plugging processes. Traditional acid treatments were attempted together with the automatic replacement of badly affected wells. While full production was specified to be 1 million gallons per day, historic data indicates that this goal was rarely achieved and there were frequent system failures. Modeling projections indicated that a risk existed of discharges into the Big Black Creek itself.

In addition to the plugging problems, it was also recognized that there was a high mechanical failure of pumps. These pumps were failing after only 10% of the expected component lifespan. This may have been a result of microbially induced corrosion. It would appear, therefore, that the BN site was subject to not only severe plugging problems (which may be aerobically driven) but also corrosive problems (which would, in all probability, be anaerobically driven). In view of the pace of deterioration, it was decided to treat one well using the BCHT^{TM} process.

Bioaccumulation was a major concern as an event occurring within the slimes forming the bioplugs. For benzidine, it was found that the product water contained between 0.7 and 11.6 mg/L, while the slimes recovered from the bioplug formations contained, on average, 25 times the benzidine concentration found in the product water. At the lowest discharge, the concentration in the bioplugs was 8,000 mg/kg. This represented a concentration 400,000 times higher than that considered toxic to humans. Such a radical bioaccumulation as was observed here leads to concerns that the there may be distortions in the monitoring protocols generated by the unpredictable nature of the phenomenon. For example, in this case, if the benzidine becomes tightly accumulated in the bioplug, the benzidine will not be detected in the product water when analyzed. A false low risk assessment will result. When the plugs did slough and were collected on the pre-UV treatment filters, this material had the color and viscosity of cola soft drinks from top to bottom but had a very high benzidine content.

The treatment scenario selected for this project was the standard BCHT^{TM} treatment and was applied to the designated test well. The degree of rehabilitation achieved allowed the well to return to 100% of the original pumping capacity. So successful was this treatment that the remediation well pump rate had to be reduced so that only enough flow for the capacity of the treatment plant was produced. Ongoing monitoring of the treated well revealed that the flow continued to "improve with time". This is a common phenomenon with wells that have been subjected to radical treatments such as BCHT™; the improvements may continue for a month or two after treatment.

A preventative maintenance program has been introduced to operate on a three to four month cycle involving acetic/sulfamic acids and a wetting agent. This program has allowed the well to continue pumping at over 100% of original well capacity. The test well continues to perform at original pumping capacity and remains the subject of an ongoing preventative maintenance program.

G. MISSISSIPPI RIVER, 1993

In 1993, activity also centered on the plugging of levees along the Mississippi River. Over the period from 1936 to 1938, a series of legislative initiatives were introduced to upgrade the levees along the Mississippi River. The upgrading was desirable in order to hold back the planned higher river stages. This, in turn, created a greater potential for underseepage that, if not controlled, could cause sand "boils" which would undermine the integrity of the structures as such. Control of underseepage was designed through the controls installed along the levee that included berms, relief wells and pump stations. Most of the relief wells were installed in the early to mid-1950s.

The relief wells of the Upper Wood River Drainage and Levee District are located on the Illinois side of the Mississippi River, bounded on the upstream end by the existing Lock and Dam 26 structure at Alton, IL and on the downstream end by the mouth of the

Wood River. These relief wells all have an 8" inside diameter wood stave constructed well screen, wood riser pipes, gravel filter, sand backfill, and concrete upper backfill. The tops of the wells are protected with corrugated metal guards and fitted with valves to prevent back-flow of material into the well. Well depths range from 58 to 79 ft with the screen lengths varying from 19 to 55 ft.

Recently, there has been the construction of Melvin Price Locks and Dam that has subsequently led to a raising river stage adjacent to the Upper Wood River District. This has had a marked effect on local ground water conditions through an increase in the differential head across the levee. Unless these additional challenges are properly managed, there is a potential for a negative impact on existing nearby property and structures (e.g., flooding of basements and low areas and formation of sand "boils" during floods). It is expected that some, if not all, of the relief wells in the District upstream will experience nearly constant flows as a result of the rising river stage. To compensate for the additional stress level, improved management practices were actively sought to ensure the integrity of the system. In view of the bacterially induced plugging, it was decided to evaluate BCHT^{TM} treatment using the standard recommended.

The form of the plug was first ascertained (see Table Eight) and initial evidence indicating a bacterially catalyzed biofouling was evident through the formation of reddish iron oxide staining on the concrete pad around the well discharge. Downhole video inspection of the boreholes determined that there were copious amounts of filamentous slimy material hanging suspended in the well. When the well was pumped, these filamentous slime-threads remained suspended in the water showing that they were tightly attached. Microbiological testing was undertaken on the relief wells in December of 1989 to determine bacterial loadings. Waters from the wells generally had a pH of 7.2 to 7.8, while the redox ranged from -40 to +220 mV. From the population data, the wells could be characterized into different groups (this information is also presented in Table Nine):

- *Group 1*, high SRB population, few fermenters, no facultative anaerobes or heterotrophs - RW-18.
- *Group 2*, low SRB, few fermenters and heterotrophs and a small population of facultative anaerobes - RW-16, 21, 22.
- *Group 3*, high heterotrophs, moderate fermenters, few facultative anaerobes and no SRB - RW-30.
- *Group 4*, moderate facultative anaerobes, few heterotrophs, moderate fermenters and no SRB - RW-29, 34, 36, 46.

- *Group 5*, high fermenters, moderate-high facultative anaerobes, few heterotrophs, no SRB - RW-44, 92.

Table Nine
Dominant Bacterial Groups in the Slime-Rich Plugs in the Wood River Levee

	SRB	FR	FAN	HET
Group 1	+++	+	-	-
Group 2	+	+	+	±
Group 3	-	++	+	+++
Group 4	-	++	+	+
Group 5	-	+++	++	+

Note that the codes used in the above table are: FR are the fermenters, FAN are the facultative anaerobes, HET are the heterotrophs while the -, ±, +, ++, +++ represent the relative population size recovered.

Stratification was also observed in the wells when sampling was undertaken from 3 ft from the static level, midpoint (RW-22 only), and 3 ft above the bottom of the well. Six wells were included in this survey (RW-22, 18, 30, 56x, 81x, and 92). Log cfu/ml of bacteria were recorded using the standard spreadplate technique. The results are recorded in Table Ten.

Table Ten
Bacterial Populations (log cfu/ml) recovered from the Wells on the Wood River Levee

	TSOY		FERM		HACF	
	Top	Bottom	Top	Bottom	Top	Bottom
Well#						
22	4.2	4.3	3.4	4.5	3.2	3.2
18	5.2	3.4	3.7	2.5	5.3	3.3
30	5.2	2.2	4.3	3.6	5.4	3.7
56x	5.3	2.2	4.3	4.4	5.4	4.3
81x	4.7	3.3	3.4	4.4	4.6	3.5
92	4.3	4.6	3.6	3.8	4.7	4.3

Note that where the populations are assessed as the log of the colony forming units/ml growing as aerotolerant heterotrophs (TSOY), fermenters (FERM) and oligotrophic bacteria (HACF).

In general, higher populations were recorded in the upper strata of the well, with the exception of the fermenters in RW-81x where the higher population was located in the deeper strata. From the microbiological studies which included the BART™ biodetector testing, there would appear to be a major bacterial fouling occurring which was different from one well to another in terms of population, with the wells appearing to be stratified with higher populations in the upper parts of the well. The BART™ water testers indicated that there were aggressive populations of IRB, SRB and SLYM formers in all of the wells. Given the evidence presented, there would appear to be a bacterially driven bioplugging occurring

Prior to redevelopment work on the wells, pump testing revealed that there had been declines from the original specific capacities down to 60%. Numerical values for these initial specific capacities ranged from 123 to 324 gpm/ft. Because of these wide variations in the baseline values, relevant test results are recorded as percentages of the original installation values on a well by well basis.

There was a phased redevelopment planned for these wells in order to generate a better understanding of the BCHT™ treatment benefits and limitations. These phases were:

1) Thirty-two wells were tested and a program of redevelopment proceeded using TSP, HTH and surging. Chemical amounts were deliberately varied to determine optimal efficiencies.
2) Twenty-eight wells previously treated in phase one, were subjected to the BCHT™ treatment. Bacterial and chemical characteristics were determined along with pump testing.
3) Fifty-nine wells not included in the earlier phases (1 and 2 above) were pump tested and prioritized for redevelopment. These wells were treated with TSP and HTH along with intermittent pumping under a rigorous regime.

a. Phase One Treatment

After the initial pump test, the wells (except those at above the 80% specific capacity criterion) were treated with a mixture of HTH and TSP. The amounts used were based upon the well water volume and were administered in a ratio of 5 to 10 lbs for the TSP and 2 to 3 lbs for the HTH for every 100 gal of well water. The mixture was pumped into the well from a mixing barrel and further agitated downhole with a surge block. One well was surged without the addition of the chemicals to act as a control.

It was considered desirable to follow the treatment with a waiting period of 24 to 36 hrs to allow the TSP to act as a dispersing agent and assist in separating clay particles and other materials inside the filter. In practice, the waiting period was deliberately varied from 2 to 36 hrs in order to determine the relationship between the length of the waiting period and the effectiveness of the treatment. Thirty-six hours was taken to be the maximum waiting period because it is generally believed that the sodium hypochlorite (in the HTH) can only remain effective against bacterial growth for that time period. Surge rates were also varied, although the standard specifications called for a rate of travel of between 1.5 and 2.0 ft/sec. In most wells the surge rate was increased to 2.5 ft/sec but Well 35 was radically surged at 5 ft/sec.

Two wells (29 and 22x) were also treated with commercially marketed sulfamic acid product in an effort to improve well restoration. In the case of Well 29, the acid was mixed in with two of the cycles of surging. After a waiting period of 18 hrs, the well was again surged twice. Upon pump testing, the resultant improvement in the specific capacity was only 4% and that would have to be considered poor. Well 22x followed a similar treatment but with ten surge cycles after the waiting period. In this case, there was a 6% improvement in the specific capacity.

During this (treatment procedure) phase, intermediate pump tests were conducted to determine with more precision the major stages in a successful treatment. Additionally, visual observations were conducted on the state of the treatment by viewing through a clear plastic tube connected to the intake of the centrifugal pump. Grain sizes, general content, and organic residues were recorded as possible qualitative means for evaluating the success of a treatment.

b. Phase Two Treatment

Given the recoveries observed using the TSP/HTH surge technique applied in phase one, it was decided to proceed to phase two. This would involve the application of the BCHT^{TM} technique. During phase one, two wells (44X and 46) were subjected to this technique and achieved a 12% and 14% improvement respectively. This improvement was beyond the limits of the phase one technique achieved through the direct application of the TSP/HTH surge technique. The triphasic treatment system used on these wells is listed below under the shock, disrupt and disperse stages are described.

- *Shock* stage consisted of batching gaseous chlorine at concentrations exceeding 800 ppm and injecting the equivalent of three well volumes into the well. There was a waiting period of (minimally) 24 hrs in order to allow the "chlorine shock" effect. Gaseous chlorine was selected because of the much lower risk of any of the components in the disinfectant (e.g., dry chlorine compounds) reacting and settling out in the water. This would reduce the effectiveness of the treatment.
- *Disruption* of the bioplug formations occurred in this stage through the application of heat, together with a mixture of polymer, acid and chlorine. After the well had been sealed, a steam generator was used to elevate the well water temperature to 120°F by internally mixing the well water with the hot chemical mixture. The precise concentrations and selection of the type of polymer was customized in response to batch field site tests. The polymer, acid and chlorine mixture had a pH of between 1 and 2 and was injected as a hot solution at 200°F using a x1.5 well volume for the heated solution. Nozzle pressures were maintained at 500 psi, but this pressure rapidly was dissipated within the well. This resulted in lower overall pressure at the target sites (the exposed bioplugs). After the injection of the heated mixture of chemicals, there was a 48 hrs. waiting period. Once this waiting period was complete, the well was subjected to a second shock treatment as described above but with a reduced waiting period of 24 hrs.
- *Dispersion* stage followed a very similar format to the manner used in the TSP/HTH surge technique used with the phase one wells. Surging consisted of nine cycles (135 passes) of surging the full length of the well screen at a rate of 2.5 ft/sec. Surging produced between 0.2 to 1.1 ft of infiltrate coming into the wells from the collapsed bioplug formations. This material was removed from the well by pumping for 30 to 40 min using the same centrifugal-type pump as used for pump testing. Drawdown information was treated as an alternative to the pump test routine. To aid in the redevelopment, there were a further nine cycles of surging followed by another short pumping period. To conclude the treatment, an "air lift" was used to remove the remaining infiltrate from the well, pumping compressed air to the bottom of the well did this. A final pump test was now undertaken to determine the total amount of improvement.

c. Phase Three Treatment

The time frame between the phase two and phase three activities at the coffer dams of the Melvin Price Locks and Dam construction provided a useful "window of opportunity" to redevelop and test the remaining relief wells without the impact of ongoing dewatering processes. Each well was subjected to a pump test for a minimum of two hours (as in previous cases) and, on the basis of the data, a prioritization of the wells most suitable for redevelopment was undertaken. Redevelopment of these wells used a mixture of 5 lbs of TSP and 1 lb of HTH per 100 gals of water and the surrounding filter. The mixture was dissolved in water and pumped into each well using the surge block. The waiting period during which chemical action occurs against the bioplug formations was 24 to 36 hrs. Surging was performed (using small rubber-tired cranes) at rates of between 1.5 to 3.0 ft/sec. Three periods of surging were performed each with 45 passages up and down the length of the well screen. Infiltrate accumulation was measured after each period of surging, if deemed necessary. Following the last surging, the wells were intermittently pumped for one hour. This consisted of alternately pumping and not pumping in 15 to 30 sec cycles with a rate of flow of 400 gal/min.

The degree of rehabilitation achieved was evaluated for each of the three phases. In phase one, the mean original specific capacity of the wells was 139 gal/min/ft. Prior to treatment, this mean value had dropped by 39% to 84 gal/min/ft. Following treatment, there was a recovery of 15% to give a mean specific capacity of 105 gal/min/ft. The length of the number of cycles of surging was evaluated, and it was found that there was a diminishing return from increasing the number of cycles used. Maximum gains were obtained in the first eight cycles. Beyond those first eight cycles, the effort level involved did not yield a good percent improvement per cycle.

The initial test results did indicate a decline in specific capacity compared to the installation test data. However, ground water levels during the first phase of the work were higher than during the installation by between 4 to 6 ft. Because of this change, the specific capacities during the first phase may have been lowered by this difference in hydrostatic head. This could have affected the original (installation) specific capacities by as much as 25% in addition to any decline in well efficiency. Comparisons are, therefore, questionable because of these water head changes over time. Initial pump tests showed the wells to have an average specific capacity of 85 gal/min/ft with a range from 42 and 148 gal/min/ft. The standard deviation about the mean was 27 gal/min/ft prior to treatment by the phase one methodology. After treatment (Table Eleven), the percentage improvements gave an average value for the specific capacity of 106 gal/min/ft with a range of values between 62 and 173 gal/min/ft and a standard deviation of 26 gal/min/ft.

Table Eleven
Impact of the Treatments on the Specific Capacities of the Wood River Wells

	Specific Capacity (gal/min/ft)		
	Before	After	% shift
Average Specific Capacity	85	106	+24%
Minimum Capacity	42	62	+47%
Maximum Capacity	148	173	+16%
Standard Deviation	27	26	- 4%

Note that before and after refers to specific capacities obtained prior to (before) and following (after) the treatments as described above had been applied to the Wood River Wells.

There was clearly an improvement in the specific capacities with an average of around a 20-gal/min/ft improvement in all well types. Microbiological investigations that were conducted on some of the wells all indicated aggressive populations of pseudomonads (>3.0 log cfu/ml), IRB (>3.0 log cfu/ml) and SRB (>2.6 - 3.0 log cfu/ml).

Redevelopment.
Twenty-eight wells were redeveloped using BCHT^{TM} methodologies. Initial pump tests showed that the selected wells for redevelopment had an average specific capacity of 133 gal/min/ft with a range from 69 and 234 gal/min/ft. The standard deviation about the mean was 39 gal/min/ft prior to treatment by the phase one methodology. After treatment (Table Twelve), the average value for the specific capacity was 147 gal/min/ft with a range of values between 80 and 229 gal/min/ft and a standard deviation of 39 gal/min/ft.

This phase (redevelopment) was an additional challenge, since the wells had already been improved in phase one. These improvements were most marked in wells having lower specific capacities. For the wells generating a maximal capacity, there was no significant improvement.

Table Twelve
Impact of the Redevelopment Treatments (phase two) on the Specific Capacities of the Wood River Wells

	Specific Capacity (gal/min/ft)		
	Before	After	%shift
Average Specific Capacity	133	147	+10%
Minimum Capacity	69	80	+15%
Maximum Capacity	234	229	-3%
Standard Deviation	39	33	-16%

The wells used in this survey that were tested exhibited aggressive populations of IRB, SRB and pseudomonad bacteria using standard microbiological and BART^{TM} methodologies. In addition, the wells showed stratification which may become a valuable positive indicator for biofouling in a well. The premise here is that wells that are severely biofouled generate a series of bacterial plates within the well water column. These are clouded zones of bacterial biocolloids that float at specific depths within the well. This makes a real challenge for sampling from a fixed depth. If the sampling device extracts water from one of these bacterial layers, a high population of bacteria will be detected. If the sampler extracts water from between these layers, then a low bacterial population may be detected.

Final Developments.
Initial pump tests showed the selected wells included in this phase had an average specific capacity of 76 gal/min/ft with a range from 45 and 117 gal/min/ft. The standard deviation about the mean was 23 gal/min/ft prior to treatment by the phase one methodology. After treatment (Table Thirteen), the average value for the specific capacity was 78 gal/min/ft with a range of values between 51 and 137 gal/min/ft and a standard deviation of 21 gal/min/ft. Percentage improvements are listed in the table below.

Some improvements can be seen in the wells having the poorest capacities, with a reducing return for the wells that had the higher capacities. Like all previous treatments, there was a slight reduction in the standard deviation values.

94 Well Treatment

Table Thirteen
Impact of the phase three treatments on the specific capacities of the Wood River Wells

	Specific Capacity (gal/min/ft)		
	Before	After	% shift
Average Specific Capacity	76	78	+2%
Minimum Capacity	45	51	+13%
Maximum Capacity	117	137	+17%
Standard Deviation	23	21	-9%

H. CONCLUSIONS, 1988 TO 1993

Some general observations on the value of treatments and procedures can be made on the basis of the data gathered so far. These are:

a. Initial treatment with TSP/HTH appears to be beneficial, provided there is an adequate treatment exposure time of at least 24 hrs. Repeat treatments did not yield such significant benefits. The BCHT^{TM} treatment as applied by ARCC Inc. had the added (synergistic) impact of combining heat, a lowered pH, and an efficient wetting agent to make the treatment (shock, disrupt, disperse) process more successful.

b. Surging of the wells appears to be a very essential part of the treatment process with better recoveries recorded with the greater amount of surging. Recoveries are, however, not linear to the degree of surging applied. There was a diminishing return with prolonged surging, particularly observed after eight surging cycles. Chemical additions during surging were found to improve efficiency through the additional disruption of the bioplug formations. This also helped to reduce particle size and allow the separation of debris from the filter material.

c. Surging was found to be predominantly lateral in its effect. "Rebound" oscillations (echoing) were commonly observed in neighboring wells. Water levels in these wells affected by the "echo" commonly shifted approximately 0.1 in. in harmony with the frequency of oscillations in the well being surged.

d. In the field, rates of surging at 5.5 ft/sec often caused small amounts of very coarse and fine sand to infiltrate the well. Surge rates of between 2 and 3 ft/sec were found to be very effective and did not compromise (damage) the well in any way. Surge rates did not appear to be as significant as the amount of surging.

e. The BCHT™ redevelopment methods were effective and time efficient. The first two wells treated (44X and 46) improved from 69 and 58% to 83 and 70% of the original specific capacity, respectively. This is considered successful particularly because these wells had already been treated (phase one) to an "assumed point of diminishing return" with the more conventional treatment.

f. In applying the BCHT™, an advantage gained was the additional amount of biological and chemical testing performed routinely. It was found to be valuable for diagnostic testing. The effectiveness of the blended chemical and heat strategies in a triphasic treatment scenario was believed to generate a more effective control of the bioplugging problems.

g. Given that there remains experiential development in the BCHT™ treatment as performed by ARCC Inc., it is probably safe to assume these will be ongoing. It can be expected that as the process improves in equipment configurations, chemical application scenarios, refining the sequential management steps, and simplifying the monitoring procedures, the process and its application will become a standard management strategy to keep water wells sustainable.

h. From the bacterial monitoring, it is evident that the deleterious bacterial populations associated with the wells can be drastically reduced (if not permanently controlled) by the application of timely, effective, and appropriate treatments. However, these reductions tend to be short-lived and it can normally be expected that the bacteria will grow back into the well and again cause bioplugging. Routine Preventative Maintenance of the wells should be automatically put into place for all new wells to retard the rate at which bioplugging and other biofouling returns to the well. It has to be remembered that a water well is never a sterile place even after treatments, and there will always be survivors. A major strategy has to always be to slow down the rate of the return of the survivors to again biofoul the well.

i. The results of both BCHT™ and previous well water treatment studies reinforce the conclusion that bioplugging is a universal factor in the decline of well performance over time.

By 1994, the BCHT™ process had been developed to the point that it could now be applied in a realistic manner to reverse plugging problems in water wells. In the three subsequent years, major improvements were made to the chemicals and the equipment to deliver the triphasic treatment within the water well environment.

3

REVISIONS AND RETROFITTING, BRINGING DOWN THE COST

I. INTRODUCTION

In the four years 1994 to 1997, major changes included the introduction of a more effective wetting agent (ARCCsperse, CB4), which is more effective at much lower concentrations than the PM-30. Simplification of the equipment to inject the hot chemical solutions downhole and shorter over-borehole times had also been introduced. To meet the needs of the individual well owner on a limited budget, a modification of the BCHT™ process is now being developed in Canada. This technique uses a much more "elastic" pH strategy called the "flip-flop". Here, the pH is lowered to less than 3.0 (acid) and then raised to higher than 10.5 (base). It has been found that the plug formations become considerably more disrupted when this "flip-flop" technique is used. The process is know as the ultra-acid-base (U-A-B™) and has been jointly developed by Droycon Bioconcepts Inc. and Canada Agriculture, Prairie Farm Rehabilitation Administration – Technical Services (PFRA-TS). Reports on these activities are listed in the selected bibliography.

Some more recent examples of applying BCHT™ will now be addressed as specific examples of the problems involving bio-plugging to which control can be achieved. One example is a Superfund site in Michigan. It is interesting to observe the relative thickness of reports, and the (sometimes inverse) relationship of the thickness to the significance. Here, the plan specifications for the problem wells were a modest three pages long, while the specifications for the perimeter fence around the property were inches thick.
These wells were double gravel packed. The design for these wells allowed for over 2000 gallons per minute of production. Designed requirements called for 100-150 gpm. To meet that need, a 150-200 gpm pump was installed down towards the bottom of the screen. Development of that well consisted of running the pump until the turbidity (measured in NTU) dropped to less than 2, the well was now considered developed. In reality, this meant that all the

development had done was the influence of that pump sucking on the formation. Within 6 months of operations, these wells were closing up. The reason was that the wells had not been developed completely and had not pulled all of the materials out of the formation through the gravel pack that could subsequently plug the well. All of these perched materials around the wells formed a natural focus site for plugging. It is essential to invest in a proper development of the well, so that once routine pumping begins, entrance velocities can be reduced. If the well had been developed properly, there would not be so much perching of materials around the well to cause compensation with higher entrance velocities.

The bottom line is that a water well of any type should be viewed as a sustainable investment. To do that, there has to be a sufficient effort at well development so that the entrance velocities can be reduced. The customer deserves a long-term operating well; time needs to be spent on the final development phases of a well before it comes on-line.

There are two different development methods with which I have had experience and come to rely on. The choice depends on where the well has been drilled and with what material it is constructed. For the wooden wells or the double gravel pack wells and plastic wells, where there is uncertainty as to how they were installed, the best technique I have found is the old-fashioned double surge block. Where the wells were double gravel packed, had large slotted screens, or used a brick well pack, then the surge blocks would be designed to remain somewhere between half and one inch off the wall with the rubber disks. Usually, these disks would have somewhere between 1 and 3 inches of rubber at both the top and bottom. Larger slot screens would have a limitation that the strokes should not exceed 2 to 3 feet per second.

This is an occasion when the good old traditional cable tool comes into its own, that is, unless the operator gets carried away working the hydraulic levers! It is the stroke speed applied to the surge block, moving up and down inside the well, that makes it effective at removing material debris. When you push the surge block down, water is forced into going out into the formation; when the surge block is reversed to come up the well, the material debris is pulled back into the well. This push-pull action acts to loosen the debris up and bring it into the well. It is important to get that debris out of the well using some form of pump or bailer. Experience has found that bailing is not as efficient with severe plugging, because the turbulence which is created ends up putting the debris back into suspension and not so recoverable.

Many drillers are able to achieve recoveries with the use of the air-lift tube. In a good well, more can be achieved within half an hour using an air hose than you can sometimes achieve in 2 or 3 hours using the surge and bail methods. Simply by placing an air-line into most wells using the casing as a conductor, effective development can be achieved. With air-lifting, it is good to start just above the top of the screen, so that air-lifting brings the water to the top of the well itself if we have enough head space. Once the water level is up as high as possible on the air-lift, shutting the air off causes the water to fall back to, and through, the slots on the well screen. This should be repeated a number times making sure that you are watching what you are bringing up each time.

If there is one lesson to be learned about well rehabilitation, it is that about 5% of the technology is science and the other 95% is pure art. The more that an operator gets involved in water rehabilitation, the more there develops an intuitive sense of how to rehabilitate each specific well. Perhaps, the lessons are that "no one size fits all" and "each well is unique." Even in drilling wells, there is that same feeling. Although the contract calls for drilling a field of wells, each well site may tell a different story and you have to be there to learn the meaning. For example, you can know exactly what you are drilling into just by the tension coming back up to the rig. Just by the feel of that tension it is possible to know how well the drilling is proceeding and what sort of formations the drill is going through. Talking in the union halls to laborers and operators on the application of the equipment at contracted job sites for the U.S. Army's Corps of Engineers and the U.S. EPA has always shown me how intuitive they are; just watching a driller at the rig, feeling what is going on by watching the way the cable line is flexing. That activity reflects what is going on with the rig. Those operators know by the "feel" of the rig, and I can see that as I stand by the driller's side.

Air lifting is really blowing balls of air surrounded by water back out into the formation and also pushing the water up and out of the well. When the air is shut off, all of that energy moving the water again collapses with water falling back into the well and out into the formation with a lot of turbulence. Everything becomes loosened up, even the gravel and sand. After a number of surges, one final technique is to open the air-line up and let it blow while capping the well off. Typically, with this development, the impact starts at the top of the screen and works its way to the bottom.

One common question is, "For how long do we develop a well?" Again, a simple answer would be "until it's done!" One of the delinquencies is that the development specifications are left out of the plans. A common practice is to over-design a well so that it has a much greater production capability than is required. Hence, even if it is poorly developed, it will still meet the more modest production goals placed upon it rather than its true capability. For example, you need a well that will produce 500 gallons a minute. If this is the requirement, then it is common practice not to develop a well that gives a designed 500 gallons a minute. Conservative over-compensation in an order of magnitude or more is commonly called for. This means the well has a greater production capacity than what is actually required. Often, this is done by having enough screen in the well to give you far more flow than is needed due to the greater screened area. A pump test may show a specific capacity of 50 gallons per minute per foot in the formation that may have, actually, been good for 150 gallons per minute per foot. One goal that must always be remembered is the advantages of achieving linear flow into the well. To do this, the entrance velocities have to be kept down. At the same time, turbulence is reduced and the well would be less likely to plug. Development is a valuable investment in the ongoing operation of water wells in a sustainable manner.

An ideal circumstance in some cases, particularly with the larger gravel pack wells, is to employ satellite wells in the treatment process. These wells can be set (typically) north, south, east and west around the well. This gives an opportunity to move the heated treatment solutions back and forth between the well and any, or all, of the satellite wells. This is a very successful way to achieve a flushing action. By monitoring the impact of the added chemical into one well allows the other wells to be used to monitor where that chemical is going. For example, when adding acid the pH of the water will drop. When the pumping action is pulsed for short periods of time (e.g., 5 to 10 seconds), this causes an oscillation in the water, which gradually pulses that water out into the formation towards the other wells. Changes in the pH of the water in those other wells show the direction and intensity of that treatment. Clearly, monitoring the temperature at the same time will act to validate those observations because the temperature in the affected wells should also be shifting at the same time as the pH. By pushing the water up in the column and allowing it to drop and moving it slowly that way, it is possible to see that one long drop in pH coming into the well.

Once the dynamics of the treatment process has been observed, it is now possible to focus the treatment into the zones where there is evidence of plugging. This technique has been applied using hot or cold solutions, but the easiest is hot because that is easier to monitor and there is the added bonus of more vigorous chemistry because of the heat. This penetration of heat can happen quickly. In the case of the PVC soft slot wells, caution has to be applied. Care should be taken not to exceed 140°F. If the temperature of greater than 140°F is maintained over time, there is that potential for the plastic to destabilize and lose structural form.

There has been a lot of attention paid to the use of chlorine as a disinfectant and a lot of concern as to its role in the formation of trihalomethanes (THM). When treatments first started in the Mississippi area, there was an evaluation of the different types of chemicals for cleaning the wells. In the earlier years when I was primarily using pressure acidization, the State of Florida and the U.S. government had a very tolerant attitude to the treatment chemicals that could be put into the ground water and left there for a period of time. At this time, I found that applying chlorine gas was a great aid in the cleaning process applied to wells. Research showed that in the sands along the Florida coastline, applying approximately 200 pounds of chlorine gas under pressure was sufficient to start using water as a conduit to get the chlorine out into the formation for treatment. Beyond the injector, chlorine was found to have moved out distances of as much as 400 feet out from the well. What was surprising was that the chlorine took three days to get there! That treatment kept the well operational for an additional four years without any further treatment. At that time, the well was abandoned because a building complex covered it. Other wells similarly treated afterwards (late 1980's) showed similar improvements and are still operational today. These wells have been subjected to ongoing preventative maintenance and have not shown any significant losses in specific capacities.

When BCHT™ was first being developed as a synergistic treatment process, the application of chlorine gas was an integral part of the process. Initially, chlorine gas can be administered into a 500-gallon tank of water using a rotometer. This device injected chlorine gas from a 150- pound anhydrous chlorine gas cylinder. To meet this requirement, the rotometer was set to deliver 200 pounds of chlorine per day. Recirculation of the chlorinated water in the 500-gallon tank was done using an all plastic pump with stain-less steel sham seals. The solutions would be recirculated in this tank until desired chlorine concentration and pH were reached. In order to maintain control, the pH was used primarily as the marker point. Depending upon the job specifications, the desired pH application could be anywhere from 0 to 3 or 4 pH units. This is dependent upon the type of acid being used. For hydrochloric acid concentrations ranged from 3 to 15%, primarily with the 15% concentration being used where the required pH was less than 2. Once the desired concentration of hydrochloric or sulfamic acid was decided, then the recirculation tank would be used to mix the added acid to lower the pH. In early BCHT™ developments, a 300,000 Btu steam generator was used to apply the heat. This was done either to the water, the solutions, or to slurries (in the case of sulfamic acid).

Much of the research was rehabilitating relief wells around the base of dams. Commonly, these wells are set roughly 50 feet apart along the embankment. Just a simple inspection of all of the wells at the beginning can give valuable insights into the way in which the wells are functioning. In one example, there was a redox gradient that went from very oxidative at one end of the dam to very reductive at the opposite end. The reductive wells at the end which had very low (negative Eh) redox values, all stank due to the hydrogen sulfide being released. The waters were blackened and had high SRB populations. Meanwhile, at the oxidative end (positive Eh), the water was just "as pretty as you please." It was sparkling clear water which, when analyzed, was found to contain an almost pure culture of *Gallionella*. These could also be seen growing as filamentous slimers in one of the wells. Samples were shipped off to Bill Ghiorse but, unfortunately, the cells did not survive the shipment.

A critical part of well rehabilitation is the use of an appropriate jetting tool to maximize penetration of the chemical solutions into the formations beyond the well screens. The first jetting tool developed was designed for wells with a 6-inch casing and stainless steel screen. A drum washer was used to get good agitation during acidization. Observing the material downhole and coming out of the well determines just how much acidization is required to be effective. Clearly, the more effective the application of the acid into the formations that are plugged, then the more dispersed debris can be pulled out. From the early work, it became clear that the combination of heat and acid along with a vigorous agitation, the screen could then materially aid in the rehabilitation of the well.

From the debris collecting in the well, it is possible to gauge the effectiveness of the acidization. There is a balance between the efficiency of the heat and the chemicals that are being put into the well as the treatment and the debris material that comes out of the well. Even the concrete in well installations can be subjected to corrosion as a result of biological activity. In some cases, it is possible to actually see the concrete being eaten away due to this activity. Care has to be taken when treating concrete in a well, since this may become directly vulnerable to the chemicals used in the BCHT™ process.

A. APPLICATION OF JETTING

In the early stages of well rehabilitation, the jetting tools were suspended over the well, using a crude tripod with a boat winch to raise and lower the jetting tool or pump up and down the well. This was quickly replaced by an automated system developed around an electric boat wrench.

One of the concerns with the BCHT™ treatment was the generation of pressures through the jetting activities. This meant that there was a need to seal the wells. In Mississippi, there were many pop-off valves that had a tendency to fail. To compensate for this when the pressures got too high for their designed tolerances, low-pressure stainless steel pop-off valves were installed on the wells. Even then, the pressures building up down in the borehole was too powerful for the seals. A "bandaid" of discarded tire tread found by the road was called into service to stop the leaks. Well, when we cut that rubber down and drove it in around the seal, it worked. The lesson is to make sure the seals are able to take the sometimes extreme pressures that can be created while jetting or surging a well.

It is very important to seal the well properly. The trademark of a good driller is that they will be very concerned about the sealing of a well. The build-up of pressure during injection can reach 100 psi. That is what has to be achieved when wells are being injected and the last thing that is needed is the well water coming back up either inside or outside the casing and flowing out. Proper grouting should control flows around the outside of the well. Some of the wells the BCHT™ has been applied to have not had proper grouting. Even though these wells cost the U.S. Army's Corps of Engineers about

U.S. $40,000 to $60,000 each, there was evidence that they had not been grouted with cement all the way down to the toe ditch. Under normal biofouling conditions around these wells, there was no need to pressure treat because the acids would easily get out into the formation. Not only was there channeling up the outside of the casing, but there was also movement out through conduits in the biomass around the well. These conduits moved the treatment chemical not only up the outside casing to the toe ditch, but also up and out of the embankment walls of the dam itself. The water was following the course of least resistance out of the dam. Because of the hot solutions under pressure to move upwards and along the lines of least resistance, solutions were emerging out of the cracks in the toe ditch, and as little steam geysers along the banks of the toe ditch. Essentially, the relief wells were doing as they were designed to do and relieving pressure, only this time it was the pressures being generated by the BCHT™ treatment.

Examples of these "geyser effects" have occurred at several dam sites where the BCHT™ treatment has been applied. One example was a very old dam where the relief wells were being rehabilitated in Texas. The operators had just finished laying down a mud-clay mix in the toe ditch over the grass. One of the operators was jetting one of the relief wells when he suddenly jumped sky high because a steam geyser had erupted right behind him and flooded his rubber boots!

Air-lift surging of wells was not originally allowed because there were concerns the violent turbulence created in this operation would compromise the gravel packs. After the treatment in the disruption phase, it is common to observe the shattered biomass debris downhole. This debris can often be associated with silts, sands and/or clays depending upon the nature of the bioplug out in the formation. When a well is under artesian pressure, that pressure can simply force the disrupted debris into the borehole.

One thing always to remember when treating wells is that they may not have been properly developed in the first place. One example of this is a particular field of wells that had been able to pump continuously at about 60-80 gallons per minute per foot when first installed. Over time, these wells had degraded to an average of about 5 gallons per minute per foot (specific capacity). After BCHT™, the wells recovered to have an average specific capacity of 100 gallons per minute per foot. This recovery clearly demonstrated that these wells had not been developed fully in the first place. The initial development of these wells was actually completed by the BCHT™ treatment! One of the problems is that well development is often stopped when the desired specific capacity is met, rather than continued until the well is truly developed. These wells are in Grenada, and in 1996, they were still running at their required performance level, and a preventative maintenance program is conducted in a cycle every two years. They would never have lasted an additional two years without having to do preventative maintenance. Usually, the major component involves the application of acids to the well about every 6 months or so.

This application of acid at the Grenada well field involves mixing up a solution of sulfamic acid and then adding Arccsperse CB-4 as the penetrant at (final concentration, downhole) 30 ppm. This is jetted into the well, and a riser is installed on the top of the well to stop the flow and allow enough pressure to build up in the well. Once the acid mix is applied and jetted, the well is left to sit overnight. The next day the wells are surged with air-lift and then opened up to discharge the treated debris, which is pumped off. This is not a radical treatment technology and a team can treat 6 or 8 wells a day in a routine preventative maintenance program.

One thing that developed very quickly in the BCHT™ process application was the need for more heat and a better plumbing system (to avoid tripping on hoses and speed up the mobilization and demobilization times). By 1989, the volumes of hot solutions had increased to 1,500 gallons per well. At the 12-inch relief wells at the Brookville Dam, the boiler was upgraded from a 300,000 Btu/hr boiler to about 800,000 Btu/hr boiler. It was at the Brookville site that the U.S. Army's Corps of Engineers recognized that BCHT™ was not such a radical treatment process that could easily damage the integrity of the wells. The St. Louis district office now recognized both the applicability of more heat and the advantages inherent in air-lift surging. It was now possible to BCHT™ wells more aggressively with a greater dimension to the treatment.

One thing to note is that the concept of sequencing through the three BCHT™ phases of shock, disrupt, and disperse remain an integral part of the strategy. The wells at Brookville were a challenge to work with partly because these wells back into the ground; and below there is an 8 foot gallery where the wells discharge into the gallery and then flow into the river downstream. While an acid treatment became integrated into the first (shock) phase, there were problems with attention risers. This was not the function for which they were designed, or able, to do (i.e., plug off the well during treatment). To compensate for this, the expedient technique developed was to take the flapper tops off the top of each well and turn them around backwards. This consisted of flipping the flapper tops on their collar, taking the sealant and applying it between the studs, and bolting the collar down to seal the top of the well to get it to work.

B. SELECTION OF A SUITABLE HEAT SOURCE
The Boiler

It was found that the 300,000 BTU/hr boiler was limited to shallower wells no larger than 4 to 6-inch in diameter. The 800,000 BTU/hr boiler was a twin stack design. This meant that there was the opportunity to operate the stacks alone or together in either a parallel or series configuration. This boiler was diesel fired, with an electric power generator firing up the boiler through guns positioned in the base of the heat exchanger. The normal output on the pump for water passed through the boiler, 10 (U.S.) gallons per minute. With thermistors controlling the temperature of the product water in the unit, there was the ability to control solution temperatures from anywhere between $90°$ to up as high as $212°F$. Additional pumps were used for batch pouring the various mixtures. While the first phase used hydrochloric acid, the second phase used drums of sulfamic and either the PM-30 or later the CB-4 wetting agent. These chemicals were premixed before being heated by passing through the boiler. As the BCHT™ treatment experience progressed, it became evident that everything needed to be handled with a minimum of direct contact. This was because the solutions were hot and very aggressive and nothing should be touched by hand.

Jetting the chemical down the borehole became the next bottleneck in speeding up the delivery of the BCHT™, since the tool was raised and lowered up and down the well using a tripod positioned over the well. Each new well meant the manual installation of the tripod over that well. To speed this up and make it more convenient, a swivel crane on the truck was used in place of a tripod. Now, all that the operator needed to do was position the truck beside the well to be treated, move the crane into position with the jetting tool over the center of the well, and lower the jetting tool into the well. Jetting tools for this job were made using a 6-inch diameter piece of stainless steel that could easily be placed in the 12-inch diameter wells. This would allow a good positioning of the jetting nozzles downhole.

C. PROBLEMS: HEAT AND CHEMICAL APPLICATIONS
Wooden Wells

A major problem that had to be faced was "how do you BCHT™ wooden wells?" Wooden wells had a number of problems. First and foremost were the problems created by the nails and screws that pass right through the wood well into the well itself. These metallic inserts surrounded the well like a "crown of thorns" and, clearly any treatment runs the risk of having equipment "hang up" on these nails and screws. When the contractor got down there for the first time to rehabilitate the wells, there were severe problems with hung up jetting tools and tangled hoses. They ended up pulling out a tangled weave of hose and even the well screen came out! Another problem was that the casing had an extreme tendency to warp and twist as it went down the well. When the well was camera-logged before the start of treatments, there were "crowns of thorns" all the way down inside the screen. To compensate for these snagging intrusions into the well, the jetting tool had to be reduced to a 4-inch diameter to get through the "crown of thorn." Working the jetting tool up and down slowly through the well, at that time, we were manhandling everything to avoid getting snagged up on the screws and nails. Then it was still routine practice to seal the wells in order to get more of the chemical mix out into the formation.

To use the jetting tool, it was lowered down to the bottom of the well. Here, jetting would begin using the acid to cause the break up of the biomass. Once this had been done, the tool was brought up to the next portion of the screen and the treatment repeated (staged jetting). That way it is possible to keep working up the screen until the heat and acid have been distributed along the length of the well. This work is extremely labor-intensive. In talking about using air to develop the well, there is also the ability to remove the "crud and corruption" that is coming out as a result of the BCHT™ treatment. The use of an air compressor for the redevelopment of these wells by air-lifting turned out to be a lifesaver.

In the State of Illinois, there was a situation with two wells where nutrients and organisms were cross-contaminating wells. These heavy nutrient and biomass loadings were pouring out and entering the other well. Both wells had become potential point sources for contamination that had to be examined and carefully controlled. Each well had a large pea-sized gravel pack with a 50-60 slot well screen. BCHT™ treatment on these wells took anywhere from 2 hours to a day to complete treatment. In both of these wells, a variation on the treatment was that the jetting tool remained at the bottom of the well, except when the well was being pumped to monitor the treatment. Both were subjected to air-lift surging as a part of the development. One well was treated in 1989 and both treated by the same modified BCHT™ treatment in 1990. Improvements seen included a higher static water level and the water flowing crystal clear. With this validation of the technique, treatment was applied to the entire well field after monitoring the original two wells that had been treated for one year. At that time, only half of the wells in the field were actually still flowing, the rest were biofouled. Before treatment, the flow had been at an average of 3 gallons per minute from one well and greater than 100 gallons per minute. After treatment, the flows were equalized across the field. Not since that dam was built and the well field first installed had there ever been an equalization in flows from the wells in that field. It was relatively easy to measure the flow at the discharge channels into the side of the river. During the recent floods in Indiana, five years after the treatment, these relief wells were subjected to much stress, but continued to operate efficiently with equalized flows. Preventative maintenance is repeated every second year.

 Some wooden screened wells that were treated by BCHT™ had what are known as stabilizers. Such wells have wooden stays with baling wire wrapped around the outside. These were 10-inch diameter wells which were installed between the late '40s up until the late '50s and they are still in excellent shape. These wells were a particular challenge because they were all made of redwoods. One particular problem with these wells was created when the U.S. Army's Corps of Engineers was when the toe ditch was re-aligned. This called for "shooting" the tops of the wells. However, the contractor did that to one well per specifications and then walked along the wells with a chain saw and chopped the tops off all of the rest. Unfortunately, there was a slope across the field so that now the tops of some wells were sitting up too high and then there were other wells sitting much lower. Clearly, equalized flows cannot be achieved when the discharges from the wells are all set at different levels. To resolve this problem and return all of the wells to the same discharge height, the lower wells were all marked. On site there were two backhoes.

The operator simply went to the wells that were now set too low, tied a chain around these wooden wells, and literally, pulled them up a little out of the ground. For the wells that were now set too high, the backhoe was used to beat them down into the ground. Now, all of the wells were at the required height, but camera logging of these wells revealed another story. Many of the wells were shattered by the pulling and the pounding and looked like accordions. Some major well surgery had to be performed on these damaged wells. Getting them lined up again and putting in new screens saved all of the wells.

The lesson from this event was that often when you are in the field, quick decisions have to be made; and there is not enough time to send a request for information and mark time while all of the necessary bureaucratic decisions are rendered to accomplish the mission. I have a philosophy in these circumstances based upon the notion that "it is easier to ask for forgiveness than it is to ask for permission."

II. SUCCESSFUL REHABILITATION OF RELIEF WELLS

The Floods of 1992

One note of comfort is that the relief wells in Mississippi that were BCHT™ treated in 1989 came through the floods of 1992 with flying colors. These turned out to be the only wells in that whole series that functioned as they were designed to, that is, to relieve the excess water pressures. Barometric pressure in these wells did not rise to the top and there was no overflowing of the peizometers; the relief wells were doing their job. Not only that, these treated wells were the only wells that did have sand boils forming in the levees. These sand boils were a very common occurrence elsewhere, since the bioplugging wells were not able to handle the additional loads; and the excessive water pressures were being relieved through the sand formations.

The next generation of BCHT™ treatment trailers was developed between 1994 and 1997. The most important feature is that the equipment is now self-contained. One major feature is the crane which can swivel 360° on the truck and position itself directly over the well to allow easy and convenient treatment. This allows the jetting and surging to be conducted even when the well is really inconveniently placed. It is easy to work from the side, the corner or the back. The new boiler is a diesel fired single coil factory certified to operate at 998,000 Btu/hr. In practice, that unit has put out 1.2 million Btu/hr adding a potential 10 to 12 gallons per minute of hot water. It is essential to have a highly rated certified boiler that is recognized in every state for both operational and insurance purposes. Under no circumstances is that boiler modified.

One major change in the procedures has been going back to using dry sulfamic acid as an alternative to hydrochloric acid. The concern about the potential for hydrochloric acid to be involved in the formation of trihalomethanes (THM) along with the chlorine-based disinfectants had led to a strong move toward using acetic acid in 1993. Acetic acid performs the dual role of the disinfectant (for the shock phase) and the acid (for the shock and disruption phases). There had been concern as to how much THMs were being made during the treatment of a well. Practice found that the THM problem only became significant when a large number of wells are being treated in the same field, and three or more wells are being surged each day. This could involve the use of a million or more gallons of water that would then have to be discharged.

III. RISK ASSESSMENTS AND ENVIRONMENTAL IMPACT

BCHT™ Treatments

In 1990, there was a real concern as to the risk assessment and environmental impacts of the BCHT™ treatment of wells. Hydrochloric acid, still very much a treatment acid of practical choice, did pose a number of potential environmental problems. A search was, therefore, conducted for other acids that would be more acceptable. The prime directive was low or no THM products after the treatment. Of the acids, some of the organic acids looked potentially suitable but acetic acid gave the best performance from the points of view of effectiveness, economy and a broad spectrum activity. Acetic acid is usually a by-product of some other chemical process and is available everywhere. One concern is that the generic (98% concentration) acetic acid is very available but, because it is a by-product of some chemical process, there is no certainty as to what else would be in the acid. It was for this reason that the BCHT™ process will use reagent grade (50% concentration) acetic acid which has a certified content. Acetic acid did not give as good a cleaning or disinfectant action as could be achieved with hydrochloric acid and sodium hypochlorite. Again, the bottom line goal is to effectively clean the well to recover specific capacities and not that of reducing the biological loading on the system.

It has to be remembered that the biological challenge is always there, no matter how successfully a well has been cleaned or even sterilized. The microorganisms are going to return and they are going to biofoul that system again! An appropriate approach to the control of this (inevitable) event is to assume the inevitability of the return of the bioplugging but leave all of the surfaces clean and free of nutrients, particularly phosphates.

Another logistic problem was the mixing of the chemicals. There was a need to store water and be batch mixing separately two different chemical solutions at the same time. To do this, it was necessary to increase the flexibility in transferring and batching, the system was changed over to three 500-gallon polypropylene tanks. In this configuration, the forward 500-gallon tank contained the acetic acid. This would allow the batching of up to 1,600 gallons without having to return to the depot for more chemical supplies.

When experience had been gained with the three tank trailer system, it was found to be rather long and difficult to maneuver. The latest treatment system uses a single 1,625-gallon reinforced polypropylene farm-style tank. This single 1,600-gallon load of chemical solutions that will give at least 2 to 3 well shots before refilling the tank becomes necessary. By logistically arranging the trailer, it is possible to set this up with two rigs with the trailer moving between wells for the critical shock/disruption phase. By doing this, up to 7 wells a day could be treated within a 10-hour day. One well will be shocked and the other well is disrupted. The sequence of operations would be to go into the field with a batch of 1,600 gallons on-board, so that there was not a need to keep returning to the depot for batching.

A second crew of operators comes along behind to complete the BCHT™ with the dispersion phase and final redevelopment of the well. This particular technique using this latest format of rig and treatment has since been applied to roughly 1,750 wells. When success is considered as the wells reaching 75% of their original specific capacity, the success rate is at about 80%. This includes the wells that have been lost for reasons other than biofouling. In a lot of these cases, the wells are screwed up because of screen, casing or pump failures. Essentially, there is a loser in every well formation. This is particularly true of wells where the biofouling has matured to such an extent that there has been solidification (hardening) of the bioplug formation. Not only that but, over time, the plug can have moved so far back into the formation, and become so large that it is untreatable because of the mass and position it now has and occupies.

The latest jetting tool runs in four directions to achieve a jetting/turbulent action all around the device. Each well presents a different challenge in terms of the sizes and forms of the slots and the degree of biofouling; therefore, the nozzles are removable so that different sized nozzles can be installed. Under some circumstances (e.g., large slot and gravel pack), a larger nozzle with a bigger aperture can inject the treatment solutions under pressure a greater distance out into the area where turbulence is desired to aid the treatment. Commonly, these jetting tools are fabricated from a 4 inch diameter 36 inch long stainless steel round stalk with a 3/4 inch diameter 16 inch long stalk driven into it. The ends of the hole had 1-inch taps at either end so that it could be used to create a 150 psi surge block. This jetting tool has mainly been used with 8-inch wells. It is used as surge block at the same time, since practical experience has found that there is a net advantage of surging between each chemical treatment as well as afterward. The surge block components on the jetting tool consist of wipers. These are rubber disks of a suitable size to fit the well; for the 8" wells, these rubber discs were 2 inches in width.

In addition to a surge block and jetting tool, there was also a recirculation tube. This tube has extension to the bottom of the tube that can be changed during the disruption phase. With the jetting tool, we had to change the rubber discs appropriately to the well. There are metal discs to "sandwich" the rubbers in place. As the dispersion phase continues, these surge block discs may have to be changing back and forth to keep recovery of the dispersed debris efficient. This is another step where the "Art" dominates the "Science;" the experience of the operator is critical.

A major feature on the BCHT™ treatment truck and trailer is the mover hitch that is a "walking" beam. It is hydraulically operated (commercially available) and is essential during surging operations. This hydraulic system had to run for prolonged periods, so a cooler was installed on the back-side of the rig. This made it possible to operate the treatment for sustained periods of time at slow speed throttled down to the motor. This allowed the hydraulics to carry on functioning without the hydraulic fluid boiling over.

Other changes to the treatment truck and trailer included more versatility in the operating speed by being able to throttle back virtually to standing and up again quickly. For sand line, 302 stainless steel cables were used on the sand pool. One thousand feet of this cable was kept spooled at any time. If it was necessary to reach into a longer hole, the cable provided 1,000+ pounds of capacity to the sand line. This proved to be particularly valuable when having to work over a bridge. Now, these stainless steel 304 cables are in common use by many operators and are, in fact, standard equipment for most U.S. Army's Corps of Engineers rehabilitation projects.

The standard air compressor used on a BCHT™ is a 200 cubic foot per minute (c.f.m.) air compressor. To deliver the air to the well, a 1-inch pipe is attached on the end of a ¾-inch air-line. Depending on well size and the manner in which these wells become fouled up, the 200 c.f.m. air compressor is operated at 100 psi. This may also vary depending upon where the static water level is in the well, but this compressor can air-lift sufficiently well to get excellent developments. The dispersion from a well by air-lift can run from just a few to as much as 100 gallons per minute depending upon the intensity of the air-stream being created downhole.

The rig consists of a four-wheel-drive truck fitted with a hydraulically driven bridge and the boom pulls the treatment trailer. Unless it is a short distance, the trailer (and sometimes the truck) are simply loaded onto a flat bed and shipped to the site. This means no long hauling distances. Arrival at the site from an airport is better because operators are refreshed rather than worn out due to many hours at the wheel of the truck. Essentially, the BCHT™ treatment has become a "drop off" package with everything ready while the chemicals arrive to be stored at a local depot.

IV. BCHT™ TREATMENT

A. APPLICATIONS PROTOCOL

There is a standard approach to BCHT™ treating a series of wells in a well field:

1) Determine the form of the loss of production in the wells by conducting: (1) BART™ tests to determine whether there has been any biofouling; (2) camera-logging to determine the physical status of the well; (3) conduct pump tests to appraise the current static water level, drawdown, specific capacity,

and well efficiency (if possible); and (4) conduct some basic chemistry on the water to evaluate the most appropriate treatment chemistry for that well.

2) Using any historical data (cynics would say this is more likely hysterical data) examine changes that have occurred. One golden rule is that if the specific capacity has dropped to less than 50% of a reliable original, and there are aggressive bacteria involved in a bioplugging, the probability of getting that well easily back to the original production characteristics become diminished.

3) Decide on the chemistry and heat application rates relevant to the characterization of the well. Remember that "no one size fits all" and each well should be treated as a separate challenge and it should not be assumed that a common treatment will work on all of the wells within that same well field.

4) Generally, experience has found that it is beneficial to take two wells in the first instance and apply the above steps right through to treatment and recovery. By using these two wells, many of the common criteria can be established and recoveries demonstrated.

5) Once the demonstration has been completed and a satisfactory recovery demonstrated, then the other wells can be treated following that common strategy but with modifications for the non-compliant wells (wells that did not fit into the common pattern with the other wells in the field). The U.S. Army's Corps of Engineers now will contract local well rehabilitation companies to perform the BCHT™ treatment on the rest of the wells. To do this, a technical bulletin is prepared for the local contractors to follow and they lease the BCHT™ equipment.

6) Final evaluation of the effectiveness of the BCHT™. As an inventor of the process, it is not surprising that critical attention is paid to the recoveries achieved by the contractors as opposed to the ARCC Inc. crews. This manner of technology transfer appears to be working.

7) To illustrate (and celebrate the success of this transfer), after the last treatments contracted in 1997, a diversion was made to the local flea market where a couple of dozen of old miniature bone china hand-painted animals were purchased. These were handed out as a token of "a job well done" to

each and every crew operator on the BCHT™ treatment rigs. They were told that the animals were the only things that didn't get broken during the treatment of the wells (!).
8) Some things go wrong with any event particularly when there is a large element of training and still an element of technical uncertainty. In 1997, the contract described above lasted for more than six months. Two events worthy of recording as "please do not do this ever again" are:

- <u>Camera-logging after treatment</u> proved to be beneficial when it revealed a 48-inch wrench perched at the bottom of the well. The before treatment videos for the same well showed it to be empty, and so it did not take a "Sherlock Holmes" type of riveting problem solving to realize that it had fallen down the well during treatment. Well, 48-inch wrenches make nice seats across the casing of a well for tired operators but, on this occasion, the operator moved his seat and knocked the wrench which plummeted down the well. It took a day to get that wrench back.
- <u>The sequence for turning off the boiler</u> is critical. If this was not done in the right sequence (e.g., the boiler is still live but the steam hoses are all closed), then the boiler could develop a life of its own. It would suddenly pump so much steam down the hoses that they would fail. Every week, two or three 50 or 100-foot lengths of hose would be lost this way. BCHT™ is tiring work with long days. However, always remember to close off the boiler, and allow the hoses to cool down before closing them off.

Another problem that constantly causes problems is the water to be used for the treatment. The jetting tool can be "a real pain in the neck" to work with because the nozzles are so small that they can easily become plugged if there are any sizeable particles in the water. To compensate for this, there are two approaches in common use for BCHT™ treatment. First, use an 80 to 100 micron filter on the influent water. Second, use larger nozzles rather than the small nozzles. The choice depends, in part, upon the size of the well and the ability to penetrate beyond the slots into the formation. This is controlled to a large extent by the construction of the well.

B. HAZARDOUS WASTE SITES

Hazardous waste environments create more challenges for the BCHT™ treatment process because there is, frequently, the dual impact of high organic loadings and the application of air, oxygen or nitrates to stimulate aerobic biodegradation. In the early 1990's, some time was spent at Superfund site in Michigan. There were, at that site, three wells set across the head of the plume, containing benzidine. These wells had been designed to intercede on the plume and control its movement. Benzidine is a long-term environmental risk with cancer as a downstream probability. While it is not a volatile product, it is very aggressive and double gloves and a double suiting of impermeable material is essential to protect against possible splash-generated risks.

One thing that is often a problem with water wells is the rather informal manner in which some contractors can become involved in the industry. There was a case of a contract being let on the basis of the presentation of a licensed well driller's card that had been borrowed for the occasion! The lender came on board as a "straw partner." Now the contractor did not even have a well rig, so the wells were installed using telephone pole setting augers. There was no doubt that this guy was a "sharp dude." With no certification, the augers went down to the specified 100 feet below the SWL. The wells casing, screens and fitting were a mix of anything that would fit and maybe work. The well had 10, 20 and even 5-foot pieces literally rammed together. The plumbing was scrounged, and pieces were glued altogether. Slots were cut for the screen using a power saw. Some slots were across, others ripped at all sorts of angles. Needless to say, there were failures for which the answer was to just chop out the damage and put in a piece of pipe and cut slots, making a screen right then and there. Some slots were cut across and some were ripped up. When the time came to shoot the borehole with gravel pack, the method was to back up a mixer truck full of aggregate. There was no record of what size aggregate had been used (e.g., pea gravel?). When the contractor started to shoot the hole, the aggregate shut up open slots, shot rocks tumbled down one side, then moved around the other side like a snake trying to crawl down a hole. After the pumps had been removed, then the wells were camera-logged. The water had the color and viscosity of Coca-Cola®. Pump life expectancy was hardly one month. Acidization was being applied every two weeks to try and maintain some resemblance of flow. These intercedence wells finally failed to contain the benzidine plume that was now breaking out through the containment.

Clearly, the economy of installation and the lack of attention to sustainability was now reaping the just reward (failure). Even the piping was showing severe biofouling as a result of the biological activity within the installations. They had a break out on the plume. When the 2-inch pipes were inspected, there was not even enough of a hole in the middle of the drop pipe to move a pencil through. There were some areas where we pulled the pipe out and the hole in the middle of the black slime inside the pipe was the size of a dime. The BART™ test showed very aggressive activity by the SRB, SLYM and the IRB. When the analysis was run on both the liquid and "dry phase" of the biomass of the slime recovered from the well, the highest iron content (16% iron) was recovered from the dry phase at the top of the well. The biomass down in the well was also found to contain 25-26% nitrogen. In the liquid phase analyzed, benzidine was found to vary from 16 to 18% at the top to 20 to 30% below. The amount of benzidine in the formation ground water was about 16 ppm. This biomass formed in and around the intercedence wells had, therefore, been concentrating the benzidine through the years of sampling (bioaccumulation). The iron in the ground water of the formation was usually between 1.5 and 3 ppm depending on which well was being sampled. The formation was old palatial sand dunes on the western side of Michigan, and even though these wells were less than 2 years old, the drop pipes had to be pulled off periodically just to blow this slimy biomass out!

In hazardous waste environments, it is now understood that there are bioaccumulations of product in and around wells as a result of plugging. The microorganisms that form that biofouling event simply store up the materials that are not directly usable. Such is the case with the benzidine and the iron accumulating around these wells. The slime caking the pipes was originally black as they were pulled off-line, but the slime rapidly oxidized to a rusty-red brown color once the air got into contact with the slime.

In the case of these intercedent wells, there was a very close direct correlation between the amounts of benzidine and iron found in the slime coatings in the pipes. When investigating the cause of biofouling in and around a well, it is very important to analyze the slime coatings for both the products of concern and metals with iron and manganese being two which often dominate when the conditions are oxidative and reductive respectively. It gives a "picture" of the problems in the ground water that have to be addressed, even if the soil above probably will be "off-limits" for normal human use for the next two generations or more. One of the problems with some of the older industrial sites is that there never was any good record keeping, and there is only the vaguest ideas as to what the major products may have been fifty or seventy-five years ago.

Some of the by-products from the old manufacturing processes undertaken at the site came back to "haunt" the BCHT™ treatment process, particularly during the jetting. What was coming out of the well during jetting and the air-lift surging was foam. This foam formed into floating clouds of suds all over the ground, blew up into the trees, and the following morning the wind carried it over a local highway. This caused a panic because the trees and the road looked like they had just been coated with snow and this was on a warm summer's day! Obviously, the last operators of the factory were contacted concerning this foam. The general thought was that maybe one of the products had a foaming agent in it and this had bioaccumulated around the wells and in the pipes. There was not even a list of the chemicals that had been used at the site during manufacturing. It was then remembered that a long time ago, the site used to manufacture soaps. What is ironic is that even at a Superfund site, there remains a need to get more information on the events that previously happened. The activity was not limited to just soaps but also cleaners for industrial equipment.

The reason that all of this came to light was an anecdotal story of a time when there had been an attempt to put in oxidation ponds. When the lagoons were finished, a floating aerator was installed on the water and activated. The aerator churned so much air into the water that it produced foam that blanketed the lagoons and the area with a "snow foam". That was because the water contained such a high concentration of soaps and detergents.

C. HOT VERSUS COLD TREATMENT

The development of water well rehabilitation has shifted in focus from the simple "cold" application of chemicals (which did not give reliable recoveries) to the blended mixture of heat with chemicals in the 80s. The history now extends back twenty years or more in total but for the last ten years, the BCHT™ process has been developed and applied under a number of different circumstances

There are a number of standard scenarios that are in common use for wells. Commonly, it is late in the day when it becomes recognized that a well rehabilitation is required. Maintenance should be considered at the same time as designing of the well begins. It should be a major component during the development of that well. Water wells are a big capital investment from the user's point of view and need to be sustainable (i.e., long lasting). Preventative maintenance needs to be started immediately and that means good record keeping. Record keeping, when properly done and followed, means that any deterioration in the well will be recognized quickly. It is so often forgotten that water wells are essentially an intergenerational gift from one generation to the next.

Sustainability is the key component to making sure that wells are a reliable long-term source of water for the users. At the moment, most of the activity has been directed to using the BCHT™ process for the larger wells that are industrial and public suppliers and large irrigation wells. The U.S. Army's Corps of Engineers have been major supporters of the development of that technology. At the same time, in Canada, there is now an initiative to develop a simpler form of treatment called the acid-base-treatment (UAB™), which would be applicable to the smaller individual and rural wells. Canada Agriculture (PFRA-TS) and Droycon Bioconcepts Inc. are spearheading that initiative.

One important development that still has to happen is that a simple spiral bound waterproof records book needs to be developed. This would be for use by the field technicians, the engineer, the driller, water plant operators, and whoever needs to have it in their possession. It needs to be robust (e.g., throw it in the truck and it won't get destroyed) and easy to record the data and preventative maintenance practices. Until then, there will be a variety of ways to keep track of the many things.

V. DIAGNOSIS OF BIOFOULING WELLS

There are three elements that will commonly indicate whether the well is becoming dysfunctional. It is as simple as "**A, B, C**" since the signs related to the well's activity, bacteria or chemistry. If the water well is going to be sustainable, then all three components have to be monitored. Often, when talking to operators of wells, they will commonly say that they are keeping track of what is going on over time. It is very frustrating to go to a site to do rehabilitation work and asking for the records on the well field, the reply is: "the wells have been here for 15 years, and I (the operator) have been here for two years or five years!" This is sometimes followed by: "we never had to keep track of that stuff before."

Now that the well is failing, there is an interest in rehabilitation. If the well had been subjected to ongoing monitoring with preventative maintenance, it would have already been sustainable. But it was not in time, and so the questions that emerged to try and resolve the problems:

1) What loss of flow did you need to stop at (A)?
2) What drawdown do you need to stop at (A)?
3) What changes in bacterial activities were found in well (B)?
4) What changes have occurred in the water chemistry (C)?

By watching the gradual changes that will occur over time as the well biofouls and the bioplug grows, it is possible to project when treatments need to be done. Essentially, when A goes down and B goes up and C becomes unacceptable, those are all indicators of a degenerating well. Sometimes these changes can be radical such as happened in Waverly, Tennessee.

Waverly was the site of the first radical shift that allowed the application of both the BART™ and the BCHT™ technologies in unison to resolve a well biofouling problem. The problems began for the well field shortly after some construction work was done in the area in connection with the construction of a new bridge. This involved a considerable amount of dynamiting activity. As soon as that construction activity was completed, the wells began to give increases in iron levels. After that, and very rapidly, there were increases in the manganese levels in the water but at a faster rate than the iron. The routine monthly water samples now began to show too numerous to count (TNTC) for coliform bacteria. Clearly, there was a problem and it was tackled by the typical chlorination of the wells, and when that failed, super-chlorination. This controlled, for a while, the coliform bacterial populations but the iron and manganese concentrations carried on slowly climbing. This caused the city to have to shift from using the ground water supplies to surface waters and, of course, the treatment costs for the water went up accordingly. Taste and odor problems emerged and complaints, like the iron and manganese concentrations, now began to increase. To control the iron and manganese, it was decided to design an iron removal system based upon a green sand anthracite system, but this did not address the overburdening bacterial concern. This re-focussed the concern, and the BART™ tests were employed to test the water. From the testing, it was clear that there was not a coliform bacterial problem, but there was a major nuisance bacterial problem which was being "spearheaded" by the SRB and the IRB which were very aggressive.

Given that the traditional treatment approaches had failed and the costs of surface water treatment and the aggravation of the citizen complaints continued to increase (along with the iron and manganese!), it was decided to BCHT™ the affected wells in the field in 1988. As each well entered the dispersion phase, the water being pumped out of the treated well was the color and consistency of tomato juice. For each well, it took two days of lifting and developing and lifting and developing to clean that "tomato juice" out of the formation before the ground water would again flow crystal clear. Because the wells were so badly biofouled with a very large iron and manganese rich bioplug, it was decided to repeat the whole BCHT™ treatment a second time on each well! Even the second treatments gave another stripping of the bioplug and, again, the treated water flowed like tomato juice.

After the treatments of the wells had been completed, the water quality had returned to very acceptable levels, no coliform bacteria were detected in the submitted water samples, and the SRB and IRB BART™ tests showed that these bacteria had returned to background levels and were no longer aggressive. There was no longer a drawdown problem. This had been one of the earliest dramatic symptoms that there had been plugging occurring.

20/20 vision is always after the event but it does appear that the dynamiting related to the bridge construction had been a major factor in triggering the biofouling problems. The forces created by dynamiting must have caused additional fracturing in the unconsolidated formations around the wells. This would have had the dual effect of changing the ground water pathways to the wells (which may have allowed the admission of poorer quality waters), and caused collapsing of the bioplug formations that were already around the wells. It could be that the whole problem was simply created by the violent vibrations associated with the dynamiting which shook the biofilms loose that had been attached to the surfaces of the formation. It started to slough away into the ground water and come through carrying their iron and manganese bioaccumulates with them (hence, the high iron, oxidative, then the higher manganese, more reductive, and the growing bioburden of SRB and IRB).

After treating the wells and putting them back on-line, the green sand anthracite filter system was no longer needed. In routine testing, the water went back to the pre-dynamiting levels of less than 0.5 ppm Fe while the manganese was no longer detectable. From this experience, there are a number of important observations that can reasonably be made:

1) These iron and manganese problems had been caused by biological activities;
2) The growths of SRB and IRB around the wells had been acting as a very efficient natural attenuation filter bioaccumulating the iron and manganese out of the ground water. A natural filter system which had been there all the time, unrecognized;
3) Dynamiting had a secondary impact on the stability of the biofilms forming the bioplugging around the wells. This led to destabilization, sloughing and a rapidly degenerating water quality. Consideration should be given to including these types of impacts in environmental risk assessments since there can be very significant after-event costs;
4) The traditional attitude to the aging of water wells has to change. It is not so much that a well is aging but that the biological growths associated with the wells are maturing (and mature "organisms" are bigger!). The mindset needs to change so as to view a well as sustainable if the weeds (i.e., the biofilms) are kept in check;
5) Considering that a screen has plugged for many drillers, simply put on a longer tail pipe on the pump and lower the pump further down the well. This has been a standard answer to a lot of biofouling situations and this mindset had to change. It is not appropriate to simple say "the well is getting old, let's just drop the pump down a notch or two deeper."

BASIC RULES FOR WELL MANAGEMENT

- **Begin at the beginning**
- **Test regularly**
- **React accordingly**

Figure Eight

VI. PREVENTATIVE MAINTENANCE

The Waverly Concept

If there is one message that needs to be heard and understood, it is "begin at the beginning, test regularly, and react accordingly". Waverly was a clear example of the need to test regularly and not think so much about occasionally "cleaning the wells" but talk always of <u>maintaining</u> the wells. Begin at the beginning, on day one, there is information on whether the well has been properly developed, there are the statistics on such things as the specific capacity, static water levels, efficiencies, along with the chemistry and a little bit of biology. From the beginning, start tracking the trends in the data and performance. This was done at Waverly and they reacted accordingly; first of all, using the traditional approaches and then went into problem solving with leading-edge technologies.

In this regard, the City of Waverly was the first independent test site for the BART™ water test kits for bacteria in 1987. Analysis showed that the bacteria were the cause and the chemistry was the effect. Because of these initiatives, the biofouling problems are now under control by good maintenance practices. The "alarm system" for problems is now triggered when the time lags for the SRB and the IRB BART™ tests begin to shorten, which means that the bacteria are becoming more aggressive (again). Just as a farmer has to be always concerned about the weeds growing in a crop and taking away yield, a well operator has to be concerned about the "weeds" (biofilms) down a well that are damaging the crop down there (i.e., the ground water) and taking away yield (i.e., specific capacity).

The concepts developed at Waverly became a very applicable approach to the use of the BART™ tests. In 1993, the "Waverly Concept" was introduced in *Practical Manual for Ground Water Microbiology* published by Lewis Publishers. Both the biology (cause) and the chemistry (effect) became a part of a monthly monitoring program at Waverly. Once a month, single tests are performed using the SRB and the IRB BART testers. The operator watches daily to see how long it will take (time lag) before the tests "crack off" (go positive with a reaction). The shorter the time lag, the more aggressive the bacteria in the sample would be. The longer the time lag to "crack off," the lower the aggressivity and the less concern for biofouling. Normally, the time lags to the BART™ reactions range from 5 to 14 days with the greater the number of days, the less likelihood there would be of any problems. Real concerns arise when the BART™ tester reacts in one or two days, that would mean highly aggressive bacteria were present. If there is a short time lag, one concern is that the water sample may have contained some bacteria that sloughed from the biofilms into the water. Where a BART™ does "crack off" quickly, the standard practice is to go in and take three more consecutive samples to check whether there are highly aggressive bacteria in the water.

As a routine, at least every six months, the wells are subjected to a preventative maintenance that usually consists of acidization followed by chlorination. Sampling lines were run down the three producing wells to half way in the screen and just above the sump. This way samples could be collected from below the pumps. At least every six months, three consecutive samples are taken over a three-day period. If the BART™ tests on these samples show that the bacteria are becoming more aggressive, then the treatment process associated with preventative maintenance is started. Using the city air compressor, the acid of choice is either acetic or sulfamic and is batched to 6% along with the surfactant PM-30 (ARCC Inc., Daytona Beach, FL) which is set at a concentration relevant to the type of bacteria (IRB or SRB) and the level of aggressivity. Once mixed, the acid is pumped down into the well. While the acid is going into the well there is an air-lift being performed that slowly pushes the water up in the well and then drops it back down (by letting the air out). In these particular wells, the normal batch is at 500 gallons (of acid plus PM-30). Once the pumps have started flow down the well, the pump is shut off and the rest siphons down by gravity on into the well. The final downhole pH usually reaches less than 2. The acid is allowed to sit in the well over night with periodic surging with the pump rather than with the air compressor. The water is lifted up and dropped back through the pump. Since the chemical treatment is going on down hole, leaving the pump in-place means that it, also, is getting cleaned. The camera-logging of these wells showed that the pumps were badly slimed before treatment but were a lot cleaner afterwards. The objective is to keep the pump down the well during treatment. Pumps are usually slimy and efficiency can be lost if the pump is not operating efficiently.

If the next day, pH concentrations in the well water is greater than 4, that is not going to do any real damage to anything during discharge. Normally, such waters are run into the storm sewer. If the pH is lower, then lime or sodium bicarbonate is added to bring the pH back up above 4.0 before it is discharged to the storm sewer.

The wells are still running at acceptably high specific capacity rates and there has not had to be a BCHT™ treatment on those wells since 1988. There is an on-going discussion about coming back with the BCHT™ since it has now been so long since the wells were treated and there is a concern to make these wells sustainable. In all probability, the SRB and IRB are gradually biofouling the wells up, but the preventative maintenance is holding that in check.

VII. TARGETING THE BIOFOULING

Focussing on Treatment

An ideal method of conducting an efficient BCHT™ is to be able to treat "outside in" as well as "inside out". A big problem with well treatments is that they normally have to be "inside out". In other words, there is a natural method in which the treatments are all applied downhole into the well, into the heart of the biofouling. That may not necessarily be the most effective approach. "Outside in" means that the treatment applications can be started outside the well in the aquifer and moved in towards the well. There have been a number of occasions when satellite wells have been used both for treating the well, surging and redevelopment of the well. By using satellite wells set close to the treated well (10 to 25 feet away), it becomes possible to customize the sites for treatment based upon the knowledge (with BART™ testers) of where the biofouling occurring.

A whole series of hydraulically driven treatment scenarios are now possible when using the satellite screens, as well as the well screen itself, as potential treatment injection points. It becomes possible to add acid to the gravel pack (using the well), or place the acid outside of the gravel pack using a satellite well. Hydraulic turbulence (cleaning action) can now be created by either surging or by simply turning the well pump on and off in a routine manner. Five minutes of sitting in front of a clothes washer in a laundromat will demonstrate the type of action that results, essentially a washing motion! Typical success has been achieved with a sequence of pump on for 5 seconds and then off for 15 seconds. Looking down the well can indicate the ideal sequence because the water should just reach the head before being dropped by turning off the pump.

Grab-samples are very important when using a secondary satellite as the treatment feeder to the well. Since acidization remains a primary treatment technology, monitoring the pH will indicate how well the acid is penetrating the biofouled zone (either from the satellite well to the main well or *vice-versa*). Generally, a two-log pH change (it has changed by two full pH units) can be considered sufficient for a PM program. If a fuller BCHT™ is desired, then a four- or five-log reduction might be necessary. After adding the acid and getting the pH down to an acceptable level, the wells can be left to sit (passive) overnight with periodic surging. After this, the treated water can be pumped to waste. In practice, the use of a good surfactant such as PM-30 will enhance the action of the treatment.

Experience has given a number of concerns relating to the use of phosphates as a part of any strategy to clean wells, particularly biofouled wells. Almost inevitably, when visiting hazardous waste sites, there are two biological components that can be used to bioremediate the site. These strategies involve the use of either, a natural attenuation of the native microbial flora, or some cultured organisms that are available to seed the plant for treating the contaminants. There are usually a few limiting factors and the one that is usually dominant is phosphorus. Phosphate manufacturers can make claims as to the length of the polymeric chains on a polyphosphate, or how "glassy" the phosphate product is or how biodegradable, etc. Eventually, all phosphates can end up being hydrolyzed and broken down into phosphatic forms of food for organisms. When using phosphate products as a part of a treatment strategy, there is no guarantee that all of it will be recovered after the treatment. Any mass balance that shows that less than 100% of the applied phosphorus has really been recovered means that the well environment was "fertilized" with the phosphorus that was not recovered. It is important in cleaning a well to clean all of the surfaces. Having residual phosphorus left on those surfaces means that subsequent growth can start that much faster. It should be remembered that microbes hunger for phosphates as much as humans hunger for gold and microbes are even greedier!

VIII. HAZARDOUS WASTE SITES

Challenges for Effective Treatment

Much attention is now being paid to the biofouling of hazardous waste sites where, on many occasions, the biofouling outstrips the (designed) bioremediation and causes plant failure. On one such occasion, the United States Air Force called us in to try and find a plume of aviation fuel (AF) gas in Texas that they had literally lost. The reason for this was that their monitoring wells had plugged up with biofouling to the point that it was no longer possible to recover any of the AF plume which was floating on the ground water underneath. Using the BART™ testers, this plugging event was confirmed and, therefore, three wells were to be drilled within 50 feet of the three old (plugged) wells nearest to the last point where the plume was detected.

Once the wells were installed, the investigation began. On the first day, Dr. Mikkel (microbiologist) began to conduct the microbiology and the pump test and chemical analysis began. The BARTs™ confirmed the presence of aggressive bacteria and the wells were flowing at the specified rates for monitoring wells. The standard practice for representative sampling was to withdraw 3 to 5 volumes of water out of the well before taking the sample. It was then that it became obvious you couldn't walk 10 feet in any direction without tripping over a 2-inch piece of PVC sticking out of the ground. What had happened was that this particular AF plume had become, in reality, a site for special scientific study and, like a magnet, it had attracted its own central core of expertise. This meant the dynamics of "Let's study this damn thing to death" (or the LSTDTD syndrome) had taken over. It had already been decided that biotreatment was going to be the answer to this problem. To institute the *in situ* application of this decision, wells and test holes had converted the site into a colander. All of these "intrusion" wells had been logged and filed. Shallow wells drilled by mud rotary were everywhere. Along with the muds, a phosphatic mud dispersant product was also added to breakdown the muds and help get the muds out. The outcome of this enthusiastic drilling and development was that the whole environment was converted into a haven for aggressive aerobic bacteria. The shallow drilling had admitted far more oxygen into the ground, and the phosphatic agents used to break down and remove the muds formed a feedstock.

The total cost of the exercise was $20 million and there was an irreversible decision pathway that meant the project would go on. This LSTDTD exercise continues, the monitoring wells continue to plug, continue to bioaccumulate the AF gas, and the inevitable natural attenuation driven by the abundance of phosphorus will carry on. Sometimes, I think it sad that engineers and scientists have discovered a new "magic phrase" - natural attenuation. Natural attenuation should really be remembered to be nothing more than the fact that "God was in the biodegradation business a long time before you people discovered it". If you work with natural attenuation, it will be successful over nature's time. Co-operate with nature, understand and manage the natural events and attenuation is a probability at low cost.

IX. WELL BIOFOULING

A. HISTORICAL BACKGROUND

1986 was a banner year for the development of understandings of biofouling in aquifers. A string of events happened in that year. First, there was the IPSCO Think Tank in Regina, Canada. Second, the joint presentation to the Canadian Water Well Association in Saskatoon, Canada. Third, the first International Symposium on Biofouled Aquifers in Atlanta, Georgia sponsored by the American Water Resources Association. Fourth, a technical workshop on water well rehabilitation was held after a symposium on the topic. Unfortunately, the proceedings never got published. Ironically, ten years later, it could still be published and still contains some frontier findings.

At the Atlanta AWRA symposium, some papers were given by a major well drilling company. The head geologist was very proud of a presentation of a well cleaning practice that had apparently worked very well. Over time, however, the cleaning periods got shorter and shorter and the degree of regained specific capacity got less and less. There was a clear signal of a collapsing time scale between treating the well and it failing again. From the slides, it could be seen that the zigs (recovery) and the zags (bioplugging) were getting closer together and the zigs were getting smaller. Following the zigzag line showed that the overall specific capacity of the well was failing. Naturally, he was proud of the zigs (cleaning) over that ten-year period but why were there zags? These was considered to be symptoms of the well's aging. In this case, the recoveries were achieved using polyphosphates, some of which became resident in the well each time to help fire up the next generation of biofouling. This is another example of why caution should be used in the application of phosphatic products to biofouling systems. Yes, the recoveries are very good, but at the same time, the "fertilizer" can be left behind to support the next "crop" of biofilms."

B. THE COLIFORM PROBLEM

One concern is that often maintenance, when there is thought to be a biofouling or a coliform bacterial contamination problem, the good old stand-by is a couple of gallons of domestic bleach poured down the well. Bleach (or sodium hypochlorite) is a very powerful disinfectant that is particularly effective against the coliform bacteria. It remains not good enough to simply pour that couple of gallons down the well. That disinfectant activity is going to be localized and intense but will impact on the majority of the biofouling going on around that well. It is important to determine what the volumes of the water are that need to be treated, not just in the well but also back out in the formation. To achieve an effect, it is necessary to apply the commercial grade of bleach (12-15% sodium hypochlorite) and then dilute that disinfectant in a good clean tank to get a final strength of 2 to 5% sodium hypochlorite. Pour that diluted bleach solution down the well and use enough volume to get the disinfectant moving out beyond the well itself into the formation (e.g., five or ten times the volume of the water in the well). Let the well sit so that there can be a maximum impact of the bleach on the biofilms. Leave it as long as reasonably possible and then pump the treated water to waste. The water will flow cloudy and colored and then, eventually, will flow clear. That means that much of the destroyed biofilms and debris has come out, but not all. Remember that this is a simple preventative maintenance program and not an aggressive well treatment.

If the well remains a constant problem and treatments just to keep it going are having to be done more than once a year, then an alternate strategy should be used. Alternating between an acid treatment and the basic bleach treatment can be effective. For the acid treatment, hydroxyacetic or sulfamic acids are relatively easy to work with. For the bleach disinfectant, sodium hypochlorite is very acceptable because it is a liquid and relatively easy to use. Calcium hypochlorite has been used as an alternative but is more likely to have problems reacting with the water and is more difficult to dissolve.

There are wells in the southeastern states that regularly have laboratory reports showing that there are TNTC (too numerous to count) bacterial problems. This may have resulted from bacteria growing back in the formation and entering into the well. Sometimes it is possible to see these growths when camera-logging a well. The incandescent bulbs used to light up the well sometimes cause a fluorescent glow that is the prettiest green you would want to see! This is coming from the bacterial growths coming into the well from the formation. They were also detected using the BART™ testers.

When a more vigorous preventative maintenance program is needed, a solution of the selected acid is made up at 12% with the wetting agent (e.g., PM-30 or CB-4) and then dispensing that solution to the well. A good rule of thumb is to use a times three (x3) well volume dilution factor for mixing the acid volume measure of the well. Surging the well for 3 or 4 hours is essential to make sure that the treatment solution is getting back into the formation. Once this has been done, then the well should left overnight to allow contact between the treatment chemicals and the biofilms. When the well is surged and redeveloped, much of the acid should have become neutralized. If the acid does not come back to within one unit of the usual pH for that water, add caustic soda to bring the pH up. As a final step, a post-treatment chlorination "polishes and sanitizes" the surfaces that have been cleaned.

In some cases, there is an advantage in further stressing the bioplug by carrying out a "flip-flop." This means taking the pH down to less than 3 and holding it there to achieve enough acid contact time. This is followed by bringing the pH up to greater than 10. Here, it is held for an alkali contact time. By doing this pH shift, flip-flopping through 7 orders of magnitude, the microorganisms in the biofilms have difficulty adjusting to it and so become disrupted. Holding both contact times overnight improves the success potential for the treatment.

One major problem with well preventative maintenance and full rehabilitation is that care must be exercised that the whole length of the screen in the well is treated. The last thing that should happen is that all of the treatment and work ended up just at one or two spots. That is one of the prime directives in a BCHT™ treatment: when a well is shut down for treatment at any level, make sure that the whole well is treated. If this has not been done, then the bioplugging will grow back in, often rapidly, and the well will return to a sad state of biofouling. This is very important and, for the wells treated by BCHT™, every well treated that way (completely) remains on-line for at least three to five years before another major treatment is required.

C. GOOD RECORD KEEPING AND MONITORING PERFORMANCE

One thing to remember in water well management is that there is an ongoing need to monitor the health of the well. To do this, the following points need to be considered:
1) Keep good records because they will have to be referred to;
2) Monitor the drawdown and the static water level;
3) Test regularly for the nuisance bacteria, using BART™ or standard testing methods. Interpret the time lags and the reactions using standard guides or BART-SOFT™;
4) Know the well and what the well's performance can do;
5) To care to monitor the specific capacity. Treat it as you would blood pressure, if the well was human. Remember, once the specific capacity has fallen by more than 50%, a full recovery, no matter what level of treatment, becomes more problematic if not impossible.

There is a saying "the worse it gets, the harder it becomes to get it back." Once beyond 50% of original specific capacity, experience has found that we can count ourselves lucky if we can bring the wells back to 75 or 90% of that original; it takes a lot of hard and dedicated work to get it there.

Hazardous waste sites have introduced a new dimension to the practice of water well management. Traditionally, the major role of a well was to produce water or relieve pressure. Today, the technology has been modified to cope with the movements of polluted ground water. Now, the wells have to function under conditions that contain more organic nutrients (hazardous or not) to stimulate microbial activity. The hazardous materials can also become accumulated in the biofilms forming around the well causing:
1) A potential for false data gathering because of the impacts of bioaccumulation on the residual dissolved and suspended materials in the ground water; and
2) A high probability of faster biofouling stimulated by the higher nutrient loadings in the ground water.

Because of these sometimes unrecognized factors, wells (whether they are monitoring, injection, recovery or producing wells) have an above average level of biofouling problems. At the same time, these efforts have led to the development of a greater variety of monitoring and remediation technologies.

On some occasions, the emphasis may be placed more heavily on isolating the facility rather than on remediation. For example, more thought can be given to the design of the security fence around the wells than into the wells themselves. The wells were over-designed and had enough screen area to handle 2,000 gallons per minute when developed. But, that flow was not needed, so the pump installed had a capacity of only 250 gallons per minute (maximum) and was set at only between 2 and 4 feet off the bottom of the well. This pump could, at best, only use 12% of the capacity and it was set at the bottom of the well. The well suddenly began to fail dramatically and rehabilitation was attempted. It was recognized and the problem addressed only when the pumps broke suction.

When the wells were examined, the only place that had been developed right was by the intake of the pump, and even there it was only marginal. Because the well had so much screen and a large gravel pack, the ground water was able to move down through the gravel pack, as well as coming in from the other areas outside. There was biofouling throughout the gravel pack away from the screen that was now very difficult to get at, let alone strip off and remove.

Before even attempting BCHT™, a thorough jetting and brushing of the screen had to be done just to physically remove this material and open up the slots so that the treatment would be efficient. The concern was that these growths right in, and around, the well would create such a demand for the acid treatment that there would not be enough left to impact on the biofilms in the gravel pack. Once this well "crud" had been removed, then the BCHT™ was applied and the wells specific capacity recovered.

D. PREVENTATIVE MAINTENANCE, THE ESSENTIAL COMPONENT

Since that time, these wells have been on a regular preventative maintenance program to control these very aggressive problems. The argument now is between the designer who, naturally, claims that the well design was appropriate to the requirements, and the operators who have had to face aggravation as the wells biofouled and as the costs for rehabilitation came in. Remember, the report was only three pages long and as these wells were drilled, the sampling to confirm the final design for the well field was based on 2 or 3 test borings with samples pulled every 20 feet. There was no e-logging or anything else on the records. The whole site covers about 12 square miles and there were nine identical wells installed, each having the same problems. Each one was drilled the same way, developed using the same procedures and all biofouled in similar ways. Here, it could be argued that the bacteria formed such a complex organization that the biofilms became tissues and they became an animal! The security fence is still working without any signs of significant biofouling but that report was half an inch thick, rather than three pages.

Even when there are records, problems can still arrive "out of the blue." The unknown remains always something to guard against even when keeping logs (one of the most valuable tools there is in the world of water wells). An example is a well that had lost some production that was assumed to be biofouling. BCHT™ was applied and the shock phase started. When in the upper 20 feet or so of the screen, one of the rig operators came 'round to the back of the trailer with a look of concern: "It feels like I'm getting sand on top of the surge rubbers." This sounded serious enough to stop the treatment and inspect, so I called:

"O.K., lets shut the rig down, pull the tool up and take a look," while walking around the trailer to the well head. It only took one minute, but the tool was now locked tight in the well. It was jammed tight and could not be moved. Using a set of calipers, we went into the well and found that when the installers were bringing their surface casing back out, they had pulled the screen apart! What had been happening was that in jetting the well, the material we were putting in was coming up the gravel pack and washing it back over the top of the tool. As a result, the tool was packed in the gravel. We had to get another tool into that well to jet that cascaded material out. This literally tore the well apart.

The bottom line here is that all wells should be camera-logged as a normal part of diagnostic procedures before contracting anyone to come and rehabilitate a failing well. Partly as a result of this and many other similar instances, agencies such as the U.S. EPA and the U.S. Army's Corps of Engineers routinely do this to ensure that the rehabilitation is achievable.

When it comes to wells, sometimes it is good to make parallels. It would be foolhardy to drive a truck continually without changing the oil, checking the antifreeze in the radiator, servicing the vehicle until, finally, it goes to hell in a hand basket! Well, for a truck, there are normal operating procedures to ensure the truck has a long and reliable life. It is interesting that water wells can often cost more than that truck, and yet there is no normal operating procedure to ensure that it too will have a long and reliable life. Let us say that here is a well with a specific capacity of 18 gallons per minute per foot and then (due to biofouling) it went down to 1.6 gallons per minute per foot. If a treatment recovered that well back to 4 gallons per minute per foot, that is a great increase of 150%, but it is only a 13% increase towards its original specific capacity! It all depends how you want to interpret the data. If the truck had been capable of doing 90 mph, and now it was only capable of 15 mph, and if you found a mechanic who got the truck going up to 30 mph, would you consider that was a satisfactory repair? Naturally, you would fix the truck before it got that bad and so why not wells? They often almost seize up completely before anyone even notices.

There is no blame and there is no point in blaming anyone else. The problem stems from a severe lack of interest, more ignorance than understanding, and the industry is just beginning to recognize these needs and concerns. There needs to be more comprehensive information in the industry and there are a number of books that have recently been published such as *Practical Manual of Ground Water Microbiology* published in 1993 by Lewis Publishers. If you do not have a copy in the library and you have to deal with iron-related bacterial problems, it may be worth purchasing. Like the author, it's also pretty simple to read!

Part of our job on this maintenance is going to be educating the public about the fact that water wells are not simply inanimate objects put in the ground to draw water. It is very important to move them up on that learning curve. Here, in the Canadian prairies, the sustainable water well initiative is a clear step in the right direction. This workshop has formed a forum during which, I hope, you learned something from an ignorant country boy from Florida. I hope you can take this additional knowledge to your clients and help them with their water well problems. If, and when, they do have to get the new well, then maybe it will be possible to keep that well going for a much longer time at an acceptable level.

4

DISCUSSIONS

I. QUESTIONS AND ANSWERS

Questions were encouraged throughout the workshop and these are presented in the order asked and have not been re-arranged. The questions have been modified to make them self-explanatory. Delegates to the workshop (February 4, 1998) asked questions and George Alford gave answers. Comments are listed separately.

QUESTION:
Define what size wells you have been working with?
ANSWER:
We have worked on wells with diameters ranging anywhere from 4 inches all the way through to 24-inch wells.
QUESTION:
Does the jetting tool size and configuration vary depending on the size of the well?
ANSWER:
Absolutely, because you want to get out into the proximity. The size is proportional to the well. Again, that too will vary. One jetting tool that we use in wells is about 6 to 12 inches. It will be adjusted by changing the nozzle sizes in order to customize what we want to have take place. There is another jetting tool that we would use in 4 to 10 inch diameter wells. The jetting tools being used are always stainless steel including the tank washers because they give us that nice shot-back. The Roto-rooter® jetting tools that are available to clean out pipe by blasting them out, have about the same affect. All jetting tools fundamentally work by throwing the chemical out at an angle or pressure pushing it forward. The simple answer is yes, the jetting tool is sized to fit the well. There is no one size fits all.
QUESTION:
Do you ever treat wells with a diameter smaller than 4-inch?
ANSWER:
Yes, we do. We have just finished treating a field of 2-inch

monitoring wells. The whole issue of biofouling in monitoring wells has never been addressed. As a result, all of the data gathered from them is suspect (due to bioaccumulation and biodegradation around these wells). This is a big problem that is right now waiting to come down around people's ears.

QUESTION:
In well development do you try to jet with the air at the same time as the chemical solutions?

ANSWER:
Yes. Usually, this is most suitable on this final phase if you are comfortable with using it. I know drillers that are good cable tool drillers, and I know drillers that are good mud rotary drillers. If you have a preference about what you want to use, air jetting is one choice.

QUESTION:
What about jetting with water?

ANSWER:
Immediately that means air-lifting it. With the U.S. Army's Corps of Engineers, it is now in the standard operating procedures for their pressure relief wells on dams and levee structures. In a lot of cases, you do not have the water to do this successfully. But when there is enough water, it is an effective technique. We have used water jetting for a lot of years. The surge-spot treatment is another method that may be more attractive to you.

QUESTION:
You mentioned that you did some tests to see how far out the biofilms went from the well, what was the conclusion of that?

ANSWER:
When this work was done, we had a well with other wells strung out from it to the north, southeast and west. Dominant ground water flow was from the northeast direction. These wells were roughly 50 feet one way and about 75 to 80 feet going off towards the southwest. Observations required that we went back at 3 months, 6 months to check the well and what we saw was a steady increase in accumulation and activity in these other wells. When we went in before, the bugs (bacteria) lay out here; they were just "lazing around"; they didn't have that much food to work with (sub-minimal concentrations for growth). Once everything started moving through, then the whole community started developing and the biofouling began coming up in the surrounding formations. To control these growths there are a number of preventative measures but it must be

remembered that:

1. You have got to polish the cone of influence. This intense biofouling zone is the inner city, with the gallows, and what-have-you;
2. Outside of the cone of influence, we go here into the suburbs of biofouling;
3. Beyond that are the rural areas;
4. Totally away from the biofouling associated with the wells, we go into the hinterland; The problem stems from the question: "Who knows, and who cares anyway?"

QUESTION:

Is the degree of concentration (of biofouling and microbial activity) less at 50 feet than closer?

ANSWER:

There is a great deal more concentration in this area here (around the well) and, as you move away, it gets a little bit less and less with distance. When you get back out here (the hinterland), the bugs are primarily in the ultramicrobacterial state (essentially small sleeping cells) or they are lazily working along slowly and just surviving.

QUESTION:

Did you do a before and after treatment at the sites?

ANSWER:

Yes, we do whenever possible. We like to go in every three months ideally. I might mention that the well operators can also keep good records and do routine monitoring using tests like the BARTs™. It is also good to get a routine going with each well. For example, the pump at this particular well was turned on, it was set with a time clock. We tried to set up a typical situation. The pump would come on, run for 6 hours, off for 3 or 4, come back on and run for another 4 or 5 hours, off for 8 and then it would just cycle through those same steps over a period of time. Other than power losses and everything else, it ran 24 hours a day like that. We came back in and did the same sampling at 6 months and intended to see if there was any increase in the buildup. Because of the pulsing, the ground water was moving slower and in a very routine pulse-like manner, there is not that much movement out here and this influenced (reduced) the amount of biofouling. That is partly because there is a lot of water between here and there and the velocities were way down out in the formation.

QUESTION:
When you treat wells, did you observe how far the treatment went out?

ANSWER:
That was not possible in all cases. Sacrificial wells are a rarity, but we did have access to one well that was at the base of a dam, for a year before they were going to grout it up. It was part of a research project on the dam itself. We have been lucky enough that, at most of the sites where we work and do our research, our customers are at dam and levee structures and Superfund sites. So we have been lucky enough to have quite a few peizometers scattered around the well for monitoring purposes. For a dam or a levee, we have a differential water head over an embankment. As that water level comes up, the pressure is pushing this way against the dam. Alternately, the water is trying to come up underneath the dam. Essentially, it is trying to "pop the dam up" or set up "sand boils" that bring material out of the dam or over the levee causing the structures to fail. The way to stop that is to put in a pressure relief well. You drop the pressure this way and the excess pressures are diverted. Because of this, it was possible to pick up where the biofouling was. The pressures were relieved using a pump and putting the extracted water back on the other side of the dam.

What was good about this project was that the leaking dam site stayed as long as they were my leaks! What we were able to do with the series of wells to relieve the pressure on the dam, was to install peizometers along the toe ditch and on the embankments. The relief wells usually ranged from 25 to 100 feet apart. Because of the geometry, it became possible to find what kind of penetration (by the treatment) was reaching out into the formations.

Again, in hazardous waste sites, those of you who have been involved with them know that you cannot take too many steps without tripping over some kind of monitoring well or other. Treatment penetrations have also been observed at these sites, and it is very dependent on the formation porosities.

QUESTION:
After the penetration, can you remove the debris from all of the treated zone if you penetrated to 25 feet. Can you remove all of that destroyed detritus from 25 feet out in the formation?

ANSWER:
Yes, by pumping, removal is achievable if is inside the cone of influence. Clearly, the treatment is stressing the formation into

bringing that material in. Surging is especially good if you are using airlifting. With surge blocks, there has to be movement much further out than to just outside of the gravel pack (or in close proximity to the well). But when you go back to remove this detritus, there needs to be over-pumping in order to get that stuff moving out of the formation and coming in towards the well.

QUESTION:

In general, do you find that the plugging is broken up into small enough particles to allow these particles to move out through the pores of the aquifer into the well?

ANSWER:

Exactly, it is very important to know what the final particle size is going to be so that they can come out of the porous media without blocking off the voids.

QUESTION:

What kind of sizes are we breaking these biofilms down to?

ANSWER:

I know that some of you use laser particle counting to size the particles. In the field, I just eyeball the water for clarity and experience has taught me what to look for. Well, in general, the lower micron sizes of 3 to 10 are very desirable because these will move out of the porous formations reasonably well, provided they are not too tight. Both in the formation itself, and in the lab, tests have been conducted on the impact of BCHT™ treatment and, in particular, with CB-4. We have found that where you have got a particle, for example, that is in the 16 to 18 micron ranges, it can be broken down to the next step down to 4 to 6 microns. It appears to be a step-wise procedure down to 4 microns, that seems to be the lower limit of particle sizes before they disintegrate completely.

Particle sizes are very important when trying to control suspended particulate materials in the biofouled ground water environment. This does not devalue the use of biological markers. It has to be remembered that the treatment is not going to do away with the bacteria. As soon as you turn that pump back on, go back into service, the nutrients are going to come back into play around the well and there will be biofouling as the microorganisms come back in to "play." But if you have removed the substrates and cleaned the surfaces to which the microbes were attaching, it is going to be that much longer before the well is going to biofoul and you are going to have to treat that well again. If you maintain it before these microbes establish their huge slimy communities or build encrustations, then your well is going to last that much longer.

QUESTION:

Can biofouling occur under the dam itself? If so, does this mean that the water is traveling to some of the wells and not to others?

ANSWER:

Yes, biofouling can occur under the dams. A lot of dams have old river channels running through and under the dam structures. Over these old river formations, they are often lower and more permeable. As a result, there are more wells in these areas than would be put out in the "drier and higher" zones. Wells in these zones may be spaced further apart. What happens with the majority of water here is that it is still running through the old riverbed where the sands are much finer. For example, over time at Granada Lake, we did quite a lot of treatment development work. Now at the bottom of that lake under the silts, the floor was practically sealed with an iron pan. This would mean that, with the lake full of water, there would be a lot of weight pushing down on the lake floor (on that iron pan). This means when that lake fills up behind that dam or levee or in the river, there is that extra hydrostatic pressure. This pressure is going to try to "pop" the structures right out of the ground.

We were able to do some research during the flood of '93, and we found some wells that were actually cleaning themselves. They still were not as efficient as we would have liked to have them but the hydrostatic pressures were so great from the river (which was now over the top of levee and right up the sandbagged protective barriers). This created pressures on the relief wells. To relieve these pressures, one of the operators went down to open up the top cover. When he took the top off (they had floating check valves in them), the check valve shot up into the air. From my vantage point, I heard a loud bang, followed by a sound of surging, gushing water that began looking like tomato juice and ended up like a slimy black crud. What a sight to behold before our eyes! We stood there and watched while a complete redox gradient flushed out of the well! The pressures had been so strong around that well that it had flushed itself of the all of the stages in the biofouling process as the water pressures had been relieved. It was at this time, yet again, I regretted that I did not have a camera!

QUESTION:

Could the biofilm developing under the dam force water to go into certain areas and not into others. Would that create extra

pressure on the dam at certain areas?
ANSWER:

It does. That is why we have peizometers scattered all up and down the side to monitor differential pressures. If you drive across an earth-filled dam, you will see pieces of pipe or well boxes sticking up out of the ground. Some of these are read on a daily basis. By this means, it is possible to observe the pressure differentials which helps to focus on potential zones of intense biofouling (there would be less pressure differential inside rather than outside). That is how we know when we need to go in and do a cleaning on the relief wells. It is because we are seeing a sustained peizometric pressure rise in a certain part of the dam. I would imagine that Canada is in the same position as we are with this problem. What earthen dams we have now, they are going to be repaired but it is unlikely that anymore will be built.

QUESTION:

What is the difference between the iron pan at the bottom of a lake and an iron pan that is forming as a part of the plugging in the wells?

ANSWER:

There is no difference. Our colleague, Dr. Al Mikkel of the University of Mississippi, has a student doing that comparison right now. Also, he has looked at silts and iron pans in San Diego Harbor and is finding the same structures in the salt water.

QUESTION:

Have you found that you have to change your chemical applications according to what the chemistry of the water is?

ANSWER:

Absolutely, it is vital that you pay attention to the chemistry of what you are working with, especially in the area of hardness. Our rule of thumb is, first of all, to stay on the greener environmentally friendly side of things. The chemicals that we are working with now are acetic acid, sulfamic acid, oxalic acid and citric acid. Acetic acid (vinegar) is widely used in whatever form you are working with, whether this is hydroxyacetic or just the straight acetic in glacial form. Sulfamic acid is probably the oldest acid we have used in well rehabilitation. It has been used in boiler cleaning and everybody is familiar with it; it is a "friendly" chemical and relatively easy to use. Oxalic and citric acids are excellent for cleaning iron out of systems. They form their own chelating agents. In some cases, they also become effective biocides. These acids are good since they will lift up iron, hold it in suspension, and allow you to pump it. There is an entrepreneur in Florida who puts about a pound of oxalic acid with a little bit of sulfamic acid in a box, mixes it up and sells it for $10.00 for people to clean their driveways because of the iron stains. He said: "If they are fool enough to buy it, then I am fool enough to put it in a box for them!"

What happens with oxalic and citric acids can be a problem if you have a strong concentration (e.g., greater than 22 ppm), and there is a hardness of greater than 100 (we use 80 as our cutoff point). When the acids and the hardness are high, calcium citrate or calcium oxalates are going to form and these can literally cement the well up tight.

Another situation you need to watch when it comes to the potential reactions with hardness is when using calcium hypochlorite. For example, when chlorine pellets are dispensed down the well, do not forget that the binder is already there - the calcium. If you use HTH or another form of calcium hypochlorite, this will amount to 30%. What is happening to the other 70%? The labels will commonly say "inert material". It can make the scaling up inside more serious. I have been called in to rehabilitate wells that have been disinfected with calcium hypochlorite, and this treatment made the wells worse than they were before the treatment! These types of wells are difficult to recover, but a combination of acid (e.g., sulfamic or hydrochloric acid) along with either air or hydrofracing the wells got this cemented material busted off the surface and dissolved out of the well. I am not a big fan of hydrochloric acid because my experience is that it is messy to work with. Sulfamic acid appears to work well when it is necessary to dissolve calcium deposits.

There are many people out there who are selling "magic potions" for cleaning wells and often it is nothing more than sulfamic acid. Generally, it has been stepped down to about 30 - 40% concentration and the charge is anywhere from $3 to $20 for every pound in the bag. At the local chemical supplier, it is possible to buy 99% concentrated product for less.

The important thing to remember is to get your pH down in the water, and control that by making further additions of the acid. If you heat the product up before applying it, then it may make it more effective.

QUESTION:
How do the various alternative acids compare to hydrochloric acid in terms of corrosivity against the screen slot openings?

ANSWER:
Hydrochloric acid is very aggressive compared to sulfamic and acetic acids or the detergents. For these other acids, there may be pitting and staining, but we have yet to lose a well due to these acids "eating out" the well. Keep in mind that, if the only thing keeping the well together was the encrustations, then the acids would cause the well to collapse. That has been the case, particularly, on some of the very old shutter screen wells and the iron drill-perforated wells. Once we removed the encrustations (hardening biofilms) that was all that was left of the casing, the sand just caved the casing right in.

On another occasion, there were some wooden (redwood) screened wells. These were made like barrels with stays, except that these stays did not meet, the gap was the screen! There were soft (rot) spots between the stays. The U.S. Army's Corps of Engineers has over a thousand of these wells scattered up and down Mississippi, Ohio and Missouri. They were installed in the late '40s to mid-50s and they are still operating there today. Some of them have failed in the past, many of these were gravel packed. When the wells were rehabilitated, the operators got just a little "too vigorous" on some and literally pulled some of the stays out.

When it was decided to try BCHT™ on these wells, the first thing was to camera-log the well and it was bad. The objective was to attempt to dissolve out the biocemented zone. We went in and cleaned the well. We camera-logged the well; we knew when we had a "bad" well. We wanted to see if we could dissolve the cement, but when we went in and found the broken stays, this changed the strategy. To rehabilitate the well, it was decided to concentrate efforts on cleaning that one (biocemented) zone. When it finally collapsed, the gravel pack just flowed right in!

It is always a concern to have a well that you got a little vigorous with. On some occasions, plugging problems appear to originate above the screen level that should, therefore, also be above the gravel pack. Where this has happened, it often appears that the gravel pack has also intruded higher in the well. It may be that over time, the well is self-developing because the well was not developed properly in the first place.

QUESTION:
Are particle sizes similar close in and out from the well? During the cleaning process, will the particle sizes away from the well be too large to get broken down enough to pass through the formation and come into the well?

ANSWER:
Away from the well, generally, we have not been seeing all this biomass development (common close to the well). Usually, the biomass consists of small clusters of organisms maybe gathered around some type of concentration of organic material or metals. They may lay down some amorphous or crystallized structures incorporating iron or manganese. My thoughts are that the particle sizes would actually be rather small. Some of the particulate matter is going to be dominated by the ultramicrobacteria which have a sub-micron size scale. Another factor to take into account is that the redox gradient is going to be very shallow and probably reductive. This will not encourage the types of consortial activities that would produce sizeable biomass and particulates.

QUESTION:
What is the cost to install a 10-inch well?

ANSWER:
It will vary on the number of wells being installed and on the screen length, but just a general 10-inch well, like a relief well being placed along a river. When the contract calls for 50 to 300 of these wells to be strung at the same time, the cost will commonly range anywhere from U.S. $2,500 to U.S. $5,000 a well. A single well has a lot of variables involved and the cost could range from as little as U.S. $5,000 to as much as U.S. $30,000.

Do not forget that there are obligations to maintain even when it is going to be abandoned! This would apply even if a well is about to be abandoned. For example, over here there is an old well all messed up with iron bacteria, so a new well is going to be drilled. To do this the driller decides to move away from the old well by 100 feet. This cuts the costs of installing the utilities to the new well and has the advantages of putting the new well close to the old. That means there is a risk of cross-infection of the new well from that old biofouled well. There is a hidden cost and a very large hidden risk.

When a well is closed for abandonment, disinfection is commonly applied. The simplest practice is to add some hypochlorite. If calcium hypochlorite is used, it may not even go into solution and so will be relatively ineffective. A better practice is to chlorinate with a sodium hypochlorite solution and try and force it out into the formations by surging; adding more water to the well head and force that disinfectant back as far as possible into the formation with surging. Once this is done, seal off the well so that it isn't a conduit for oxygen to enter back into the formation and cause aerobic biofouling. If that neighboring abandoned well is not sealed and becomes an oxygen conduit, it will overlap with the cone of influence in the new well. That means that there is going to be oxygen pulled into the new well from that old "abandoned" well. The redox gradient is going to move as air is being pulled into the formation. This will increase the probability of biofouling. Another hidden cost that has to be considered.

One final point to consider is the positioning of a pump in a gravel-packed well. One cost that may not be obvious is that if the pump is set to draw water vertically down through the gravel pack as such, rather than as a laminar flow directly into the well, there is a greater probability of severe biofouling throughout that region of the gravel pack.

QUESTION:
How large can a cone of influence be around a well and what effect would that have on the positioning of biofouling in the well environment?

ANSWER:
First, the geologist, hydrologist or the driller may be able to advise on the probable size of a cone of influence. Obviously, the movement of the ground water speeds up as it enters the active region around the well influenced by the pump demand from that well. The geometry of that "cone" is also critical, since this will influence the rate of movement of the ground water above the ambient flow rates. As a rule of thumb, when there is no information, a good starting point would be a couple of hundred feet if not more, depending on that cone's geometry.

There have been a number of rehabilitation contracts when there has been overlapping biofouling situations from one well to the next. Essentially, the overlapping cones of influence cause an integration of the biofouling between the wells. This means that the biofilms forming around these neighboring wells connect and produce a much larger (and potentially more aggressive) plugging event. There have been contracts that have been voided because the clients insisted on putting the wells in too close and were not prepared to accept that there was a very serious biofouling risk which could compromise the longevity and operations of those installations. There are cases when there is no other choice and, in these cases, much more diligent preventative maintenance has to be practiced from the beginning.

QUESTION:

There are about 200,000 water wells in Western Canada that have been installed at about C$5,000 each. There are at least a billion dollars of capital investments sitting out there in those wells. How long should they last and how do we get them to do that — last a lot longer than the present (estimated) average of 10 or 15 years?

ANSWER:

Some of the answers have already been covered and preventative maintenance would be a major factor in extending the operational life span of these wells. Certainly, there is a need to understand the well field problems and, in particular, the biologically driven fouling elements that are probably the least understood and possibly the most important. Is it possible, given the numbers of wells in the Canadian prairies, to come up with the annual losses from that investment due to well failures? That is a cost to society as a whole, not just to the individual well user.

QUESTION:

Some of the work we did in Alberta prairies in just one municipal district was directed at looking at the user attitudes to operating their water wells. That survey showed something like 20 to 30% of these wells were so significantly biofouled that they could no longer produce water in the amounts they were originally designed to produce. If you add C$5,000 for each well out of the 200,000 water wells that are biofouled (e.g., 30%), that means that there is several hundred million dollars worth of water-related structures that are, in all probability, going to fail over the next 15 years. Is that to be considered substantial (e.g., C$20,000,000 losses per annum) and what kind of investment addresses the reliability of that infrastructure? What are the hard realities of the economics in the biofouling of water wells?

ANSWER:

There certainly is a problem, but it is not fair to lay all the blame on the "bugs". It is necessary to look at the "bigger picture" that is the bio-electrical-chemical-mechanical processes that are now impacting wells. Yes, the microbes are a major factor in the biofouling, but, as I like to say at conferences: "there's a lot of stuff going on down there in them those wells!" There is no doubt that it is a very dynamic situation.

The dynamics begin with the fact that you have water movement throughout the region around the well. As a result, there are chemicals moving through here, metals moving, reactions taking place and above all, you are moving from an anoxic to an oxic environment! There is precipitation and mineralization taking place. This is not commonly understood. For example, I have had people ask me: "My grandfather had a well and it lasted him all of his life and it lasted my father all of his life". To this, my standard reply is "Yeah, but grandpa and your pa were taking it out by the bucket full, and you are taking it by the (cubic) yard!"

Back when I was full-time in the drilling business, most often we would figure that 12 gallons per minute for a small household would be fine. This was reasonable because the Florida ground waters are pretty well abundant, both in shallow and deeper aquifers. In most cases, when we put a well into serve a single residence, our objective was to deliver to the house probably about 20 gallons per minute. This was because it is a pain in the neck (and elsewhere) when someone is taking a shower and somebody else flushes the commode and the "showeree" gets that uncomfortable shot of hot water. Unfortunately, we waste a lot of water simply because it's too damn cheap. It is amazing that people are willing to pay a lot of money for these juices made from concentrates to just add water to it. It is ironic that you pay a lot of money for a tablespoon of syrup, perhaps a buck, and are not willing to pay for the water coming right out of your tap.

I sometimes think that if this drilling of wells continues without adequate preventative maintenance, we are going to reach a point that there will be nowhere left to put wells. One of the major driving forces is that it usually costs three times as much for surface water treatment than ground water. For example, the City of North Battleford has an advantage in that the ground water can be treated for about a third of the price of surface water. Extracting ground water can, therefore, be economically desirable.

Wells will always be there; people in the drilling business should not see rehabilitation as taking away work from them. It is important to understand that there is economic value to you in rehabilitating wells. There is also an economic value to the community and neighbors. There is a profit that can be generated by servicing the wells and thus making them more sustainable.

COMMENT:

In Saskatchewan, we traditionally lose 1 or 2 people a year while they are acidizing wells. While there is a good group here, I would hate to see the public doing some treatment unless it was safe and they could easily handle the chemicals and the procedures.

ALFORD:

Remember, the amount I was showing you there — less than 10% concentrated acid should not directly be such a problem. Is the hydrochloric acid blowing back on them or is it the gas being generated that has caused the deaths?

COMMENT:

The problems are with the gas, hydrogen sulfide. There are a lot of people out there and they are still wanting to clean the wells for themselves — and the two do not mix.

ALFORD:

I am not advocating that the homeowner does it (chemically treats a well) unless they understand what they are doing and follow the directions scrupulously. Professionals, preferably, should do these treatments, and we lose a lot of them every year, too. For maintenance, clearly, you do not need to use such heavy concentrations of acids to remove first stage biofilms as you would need to remove mature masses. What you are also looking for is a detergent action on the maturing biomass to break it up.

COMMENT:

Pouring 4 or 5 gallon jugs of liquid bleach down a well is a lot better than just doing nothing. Even if it just treats the pumps and they were all that was disinfected; would this stop the pumps from sliming up so quickly?

ALFORD:

We have not reached the homeowner yet. What I have been addressing, primarily, is the commercial and public supply sectors. They should be minimally mandated to keep some type of records relating to the management of the ground water resources. I have gone into Superfund sites and 50 year old dam and levee sites – and I have asked: "Where are the records?" Records need to be started with all of the new wells. Maybe that is an initiative that someone has to take, that is, to develop a simple system of well record keeping that would be easy to follow. This would mean that when a city takes possession of a well, they are handed six copies of the standard record manual for the management of that particular well.

There are exceptions to every rule. There are plenty of times where the methods employed to manage the wells are possibly unsuitable. For example, there have been lot of times in the smaller systems that it is the mayor's brother-in-law (or his nephew) who runs the water plant. That is the guy who comes around to the treatment plant once a month and signs the report form. If we can start reliable record keeping on new wells, and then have all the repaired wells started on a reliable record system, you are going to end up with a much happier client.

QUESTION:

What is the minimum number of wells you would have to rehabilitate to be economical and effective on the biofouled wells. Are there any simple alternatives?

ANSWER:

In a household well, air-lift it first if you know there is an iron-related bacteria problem in the area – are there any slimes and so on? Air-lift it very quickly, just to move out any of the crud. Surging can be used for a new well. Next, add sulfamic acid or whatever acid you are used to working with, set the pump back into the well, and the next day test the pH of the water with some litmus paper. When the acid has impacted and been neutralized, the pH should come back up within an acceptable range. If it doesn't come back up, flipping the well on and off will gradually cause more (turbulent) activity and the acid will become neutralized. You have got to flip it on and off periodically and pull out any debris. When the pH comes back within range and the water clears up, turn the system back on. Remember that sulfamic acid can be purchased for a dollar a pound.

COMMENT:

That's not what we're paying for it.

ANSWER:

How are you buying it? Are you buying it as NuWell® — that is a good product. But with the NuWell® product you've got a stabilizer in there to help control it from etching on the pump. This is important if you have an iron pump or metals down there in the borehole. The product also has a pH indicator in it so that you can tell when it is spent. When you purchase that product you are also paying for all the technology that has been added to it. I have customers who have worked with it mainly because of the convenience of the pH indicator in it. It saves them a lot of hassle, they can just go back and put the pump on-line once the pH shows the actions is over and the water is back to it's usual pH.

If a client has a new well, I will go and make a super-saturated solution of the preferred chemical at the surface and then pour it down the well. This is rather than trying to flush water in pulses to get it back into the formation. My preference is to jet the solution into the formation and this works quite well. You can use a combination of jet development along with pushing your acid back through the screen.

COMMENT (Brent Keevil, graduate student working on the UAB™ treatment project):

Just a little background on North Battleford (site of a Canada Agriculture UAB™ project). The well field has a surface water recharge from the North Saskatchewan River. There has been a history of biofouling and well failures due to gradual losses in the specific capacities. These could not be readily corrected by treatment (primarily, acidization). Well #15 became a part of a pilot study to look at rehabilitation using UAB™. This well had a static water level approximately 14 feet below surface, and had 20 feet of slotted well screen from 60 to 80 feet down. The original specific capacity at the start was approximately 203 gallons per minute per foot and, according to the original logs, there was a potential to produce approximately 360 Imperial gallons per minute.

Once the well started to fail significantly, the city went through intense standard cold acid treatments over a period of 5 years. Marginal recoveries were sometimes achieved, but there remained a steady decline in specific capacity. This decline reached a point when PFRA-TS was approached, as well as Droycon Bioconcepts Inc., to try and correct the problem. The specific capacity last year dropped to 1.96 Imperial gallons per minute per foot. As a result, they were really only able to pump the well at approximately 50 to 64 Imperial gallons per minute at a drawdown of only 8 feet above the screens.

PFRA-TS and DBI cooperated to apply a UAB™ treatment to that well. After the treatment, the well could be operated at double the pretreatment capacity. Since then, it has been maintaining that specific capacity. Our concern is, of course, how to be able to keep Well #15 from getting worse again as a result of the return of biofouling. The plan of attack that we have developed now is, basically within the next month or so, they are going to go in and chlorinate the well. To do this, a chlorine solution will be prepared with approximately 3 times the volume of the water column from static water levels to the bottom of the screens. This solution will be put down the well through the pump down through the (reversed flow) turbine pump itself. It will then be allowed to sit, and then a surging will be started just by bringing the water up until it just starts pumping up at the top, and then dropping it down again. This was done for approximately 48 hours. At the end of this treatment the well was pumping clean.

There is now a need to reassess the situation. It begins to look like the gravel pack is starting to come into the screens as we develop the well. George (Alford) has mentioned the value of putting some satellite wells around and checking the subsurface hydrology. What we are going to be looking at with this well, is doing the "flip-flop" (as George would call it). This means "flip-flopping" between lowering the pH at one time (flip) and raising the pH above the normal pH of the ground water (flop). As an ongoing initiative to make that well sustainable, we are developing the North Battleford post-treatment strategy. This involves routine monitoring, particularly checking the drawdowns, pumping rates, along with the chemistry and biology (using the BART™ tests). Altogether, this means more attention to record keeping (such as pumping rates and drawdowns twice a week, and the chemistry and bacteriology monthly). Wells are a large capital investment, and like a truck, they have to be maintained.

II. SUMMARY
(by D. Roy Cullimore)

What we are seeing is that this is a very challenging time in water well rehabilitation because essentially this has been one of those "out of sight, out of mind" events where you do not actually see it and so do not need to worry about it. There is often no need created until there is a panic situation, because there is not enough water or the water quality has degenerated. You have to suddenly look at two alternatives. One is to drill a new well; the other is to rehabilitate the existing well. The traditional attitude has been to simply drill a new well. The costs are changing and "greenness" is developing and the philosophy of keeping the equipment going for longer is now very desirable. We are not quite so much of a disposable society compared to the time when wells were essentially disposable items. Once it becomes understood that it is not that easy to truly rehabilitate a well, then you have to look at what is causing the well to degenerate.

I started in 1971 with a Saskatchewan Research Council research grant to look at whether iron bacteria even existed. I was told by one of the leading professional hydrologists: "I do not know why they are wasting their time funding this project. All of these events are chemical and physical! It is ridiculous to think that bacteria could live in such extreme environments". Since then, I have been into more extreme environments than that and the bacteria are growing there, too. What is happening is the realization that water wells do degenerate. The National Water Well Association and the American Water Works Association both realize that something has to be done. The first thing that has to be done is you have to find the tools to monitor the problems and create good solutions. Going back to the Saskatchewan Research Council in the early 70s, they recognized that there was not the means and put the challenge out to me in 1971. It was not until 1986 that we came up the technique that is known as the BART™.

The BART™ is a very radical microbiological monitoring technique. It is radical in the sense that the principles are so damn obvious! It is also very simple and you can do it in the field. It goes back to some of the work that should happened in the late 1800s. Winogradsky could have thought of the BART™ but he did not; it was left for me to think of and patent it. Now the tool is out there to recognize iron bacterial loadings in water wells. What has happened now is that there has developed a greater realization of the high costs involved in failing (biofouling) water wells.

John (Lebedin) of the PFRA-TS, Canada Agriculture talked briefly about the fact that we have 200,000 water wells in the Canadian prairies. We have an investment of at least a billion dollars. How much is spent on maintaining that billion dollars of investment? How much of it is simply being left to degenerate? When you go to the American system, you magnify everything; 5,000 water wells are being drilled every day. That is an awful lot of new wells. How many of those wells are really replacements for plugged (biofouled) wells?

The American Water Well Association Research Foundation, finally, due to lobbying by some of their directors (not by me!), have decided that they would call for proposals to look at water well rehabilitation. There were 3 proposals submitted. I supported one proposal which was the Leggette, Bradshears and Graham (LBG) proposal. This is a large environmental engineering group in the United States with offices across the country. Their proposal to the

Research Foundation was for C$1 million dollars of effort. The final report is due in July 1999 which is not far away. The object is for LBG to evaluate all the water well rehabilitation techniques - all of them from BCHT™ and UAB™ through to shock chlorination and acidization. Twelve municipalities in the States have signed on and offered their resources. In-kind contributions and cash contributions lifted it up to one million dollars.

The Regina-based effort is one small part of the investment to assess and improve the techniques in water well rehabilitation. The Regina effort forms one small part of this co-operative research effort. The reason the BART™ testers were selected was not because I love these tests. It is because LBG had used them. The AWWARF knew of them and mutually agreed that this was the only test that would allow them to look for the biofouling that is occurring in water wells in a realistic manner. That is a tremendous compliment. I never realized that the BART™s had reached that state of acceptance. The challenge is to come up with a standard method. For the last two months, the computer has been glowing red-hot, as I have been trying to write a standard method to look as to whether a well is biofouling or not.

You can imagine what LBG has to go through to complete this project. How can you do a scientific experiment when there are so many different types of consolidated and unconsolidated wells at this depth, at that depth? How can you possibly prepare a guide applicable to the variety of treatments that are out there?

What we are looking at is trying to come up with the first guidelines to address water well rehabilitation. Can we lean on Europe; what about Europe? They now have a ground water commission. They, too, are struggling with the question: "what are we going to do about this lost production in ground water resources?" There is a growing global realization that the old premises: ground water is cheap; you do not have to treat it so much; and it is more economical; are not appropriate to the modern-day "greening" society.

In some areas of the States, the ground water is being sucked down until "how much more will there be?" All of the above has led to the AWWARF initiative. One thing that I have suggested is the issue of cold versus hot treatments. It is a very interesting fact that if you (as George Alford often puts it) wash dirty dishes in a sink with a detergent, you have to use hot water. You are not going to use cold water to clean the dirty dishes. Why then would you use cold water to clean a dirty well?

Another way of looking at the cold versus hot question is, if you take two sugar cubes and drop one into hot water and one in cold water, which the cube would dissolve the faster? Naturally, in the hot water! We will not get into whether hot water or cold water freezes faster in a freezer – that is another story. The bottom line there is that if you are going to rehabilitate water wells, heat makes sense because chemical reactions occur faster. There is going to be greater trauma amongst the biofouling (plugging) organisms down there in the biofilms.

Plugging is an interesting word. There was a problem with that word, which George and I created back with the U.S. Army's Corps of Engineers when we would say clogging and they were concerned and told us that we could not use the word "clog" or any of the derivative terms. Why not? Because everyone thinks that clog is a chemical or physical concretion and does not involve microbes. So, can we use plug? Yes, you can use plug. For the next 15 years, we have used the word "plug" and any of the derivative terms. What has happened is that a lot of people now talk about plugging water wells. If I say clogging, they will ask: "what do you mean?" Another common term is *bioplugging*. There's a whole set of changes in the drama of the water well management that is going on today. John (Lebedin) came up with the term *bioplume* which is actually a very nice term, and one that is very expressive along with the sibling term "bioplug."

What is going to happen next year (1998-1999) is that all these well rehabilitation techniques are going to be compared. I think that, before five to ten years ago and as a result of the sterling work that George has done, people thought it would be impossible to heat a water well and get it up to 80°C. Now, it is well understood that the potential is there.

Brent Keevill and Twyla Legault, two students who have been working with me, were in Las Vegas in September, 1997, where they gave presentations. The chairman of the session said he had never heard two students give such professional presentations. I take my hat off to them for they deserve it. The bottom line is that the discussion went on for an hour and a half afterward because of the things being discussed. Brent was casually talking about pH probes going down the well that had been heated to 80°C or 160°F. The probes were working at distances of up to 300 feet. Many people did not even know that you could put a pH meter into a well and that it could work at those depths. There was actually a lot of discussion between Brent and Cole-Parmer about how to make their pH probes more effective at long distances down hole. There was technical fallout in favor of Cole-Parmer and progress was made!

Twyla has been dealing with a survey of wells in Kneehill M.D., Alberta. Here, there was (to me) a mind-boggling situation where we had a well biofouling dominated by sulfate reducing bacteria on a scale that I had never seen before. That raises the whole question of whether leaking oil and gas wells (methane driven) could actually form a feedstock for the SRB populations around impacted local water wells. The probability appears to be that "yes, they do." If so, how do we handle that one?

With the LBG initiative the AWARFF is clearly pointing to the need for reliable applicable standard methods for water well maintenance. This is where the PFRA-TS of Agriculture Canada comes in with a mandate to get a sustainable water well initiative (SWWI). Lewis Publishers will publish this workshop as the second monograph in the sustainable water well series. The title to this series is a tribute to the initiatives of the PFRA-TS. When the SWWI was announced, it got into publications of the American Water Works Association and the National Ground Water publication; they were pleased to see this initiative happening. There has got to be a change in the attitude that a well is "a single generation disposable item" to the attitude that a well is "a multi-generation sustainable item." This could happen with proper management.

You heard George say a little bit about monitoring wells. Think of all the monitoring wells that are installed on hazardous wastes sites right around this planet. Forget North America – think about the planet, Earth. All of these wells may be subject to the risk of biofouling. What would biofouling or plugging of the monitoring wells mean? Bioaccumulation of the hazardous waste material around such wells would mean that the data gathered from water samples from such wells would be meaningless (e.g., too low – bioaccumulation; too high – sloughing of the biofilms). Are we spending billions of dollars on meaningless data? Why are we doing this? Because it is standard procedure today, the people that developed that procedure were ignorant of biological fact. The biological fact is that we humans are a small a part of the whole biological community on this planet. From the smallest microbes to the largest Sequioa tree, all life is inter-related and integrated. We are moving from a period where there have been seeds of ignorance were planted into a core of arrogance with which the human species functions. That means we knew all about water wells even when we did not know a thing about their biology. That is now changing.

One thing to remember about the biology of water wells is that it was always there. For a lot of the time, the organisms growing around that well form the bioplume that is actually acting as a water filter. George has talked briefly about that. Waverly is a perfect example of the fact that the bioplumes there were actually taking out the iron and manganese by bioaccumulation. Because it is happening naturally, it is often easy to ignore it. You should not. Instead, you should be managing it. That means more monitoring, more controls and more understanding. I am not that much of an academic that I have to have my "fix" of research grants, I actually try to avoid them unless they represent a particular challenge and are not laced with the usual dogma! However, to get into research in a cooperative manner with other disciplines is very important. I think that the growth in knowledge will now be in that direction rather than in the linear forms of science presently eulogized by some. What I hope we will be seeing in the next two decades is a dramatic increase in the knowledge base in this area.

BCHT™ has proven itself to be very successful in a variety of wells. If you look at the hazardous waste sites and the black slimes that are in those wells that the BCHT™ has got out, there is a validation of the ubiquity of biofouling. The number of places we find black slimes seems to go on forever! In the Canadian prairies, there is a growing need to develop an information base that will allow us to understand how the SWWI can be applied to the billion-dollar investment in water wells. Hopefully, we can extend a useful operating lifespan of these wells and, preferably, also to have better quality water flowing out of them. Ideally, there should be no significant levels of nitrates, pollutants, iron or manganese in the water. In a nut shell, we have got to learn to manage water wells, not just treat them as simply holes in the ground that water comes out of. There is a biological crisis down there that we have to learn to control.

III. FINAL COMMENT
(by George Alford)

The Belgians and the French did a study on the siting of pumps in relation to the screen because very often, the pumps have been sited in the casing above the screen level. When this is done, water coming into the wells usually comes in through the top 15 to 30% of the screen. Further down the screen, the water is coming in very slowly. The majority of the water entering the screened area may actually have been working its way up (or down) and through the gravel pack before coming into the well. This can be a real problem that has often gone unrecognized. What I want to do in this summary is just revisit some of the well sites that I have experience with, and illustrate just how little we know and how much there is to learn. In this case, water passing in a turbulent manner through the gravel pack in vertical rather than lateral directions is likely to extend the size of the plugging forming within the gravel pack. A laminar flow directly into the well through the gravel pack would reduce this type of biofouling.

In Poland, Dr. Wojick saw an opportunity to conduct an autopsy on some dewatering wells that were biofouling badly with IRB. The wells had to be removed to allow expansion of an iron ore strip-mining operation. Complex encrustations were recovered that reflected the pathways the ground water had been taking into the wells.

We saw forms of growth similar to that at Rocky Mountain arsenal when we had a chance to dig some injection wells. It should be remembered that water is going to take the path of least resistance, travel through the lowest pressure area until it enters a free flowing stream. If that pressure gradient is distributed across the screen area, the water will pass through into the well but, otherwise, it has to go in up along the lines of least resistance.

One way of improving flow pathways into a well is to place a laminar flow device on the bottom of the turbine pump tail pipe rather than have the standard cone strainer or another configuration of the tail pipe. This tail pipe is designed to simply allow an even drawing of water (in a laminar fashion) from the whole length of the screen. Turbines are designed to push water rather than to pull it in a vacuum. These screens on the tail pipe have a variable slot size to get even flow along the full length of the screen. Where the water flows very fast into the well, the slot sizes are small (i.e., small open area). Elsewhere where the water flow is more turgid, the slot sizes are larger (ie, bigger open area). By doing this an even (laminar) flow of water is created coming in along the full length of the screen. Using this laminar flow device has had other benefits, including much lower levels of metals precipitating on the surfaces and less gas formation due to cavitation.

The bugs we work (or are wrestling) with, for some reason, enjoy attaching and growing in high velocity areas. This is perhaps because of the food and everything being channeled in greater amounts through that zone. These bacteria do not seem to work/grow so well unless they are under a lot of stress. As they grow to form biofilms, the growths begin to slow down the flow of water and substrates. When the biofilms swell, I think that is when the "slime balls" have arrived and the specific capacity of the well declines!

In Long Island, we were having real problems with some wells. The well screens had a lot of water coming through but there was a lot of sand production. There was a focussing of water flowing in through the screens in one area, then suddenly it would move to another spot and sand would come pouring in. This appeared to be because of a very uneven flow of water into the well. I began to think getting water from these wells was like mining. You would find a rich vein (of water) and it would flow for a while and then there would be sand and another rich vein of water would open up somewhere else around the screen. Here was a case for creating a laminar flow of water into the well. When the laminar flow device was installed under the turbine, the flows into the well stopped being so turbulent and became laminar. There was no longer a serious sand problem. As a result, these wells have lasted for a number of years longer now and the practice appears to have made these wells more sustainable.

One of the risks of success is putting yourself out of business by doing things like that and then instituting a preventative maintenance program which works so well they never need to call you again! The PM primarily uses hydroxyacetic acid here and it is keeping their wells open. When they drill a new well now, it automatically goes on preventative maintenance.

Hazardous waste environments have become a major focus for my use of the BCHT™ treatment because at those sites it is not a question of whether any of the wells will plug up, but rather which well will plug up first! Rapid bioplugging, sometimes measurable in days (or even hours), has been observed at wells in the plume, in monitoring wells, in extraction wells and even wells sited on the edge of the plume, even in sundry peizometers scattered around the area. The first place that the problems often appear is on the outside edge of the fringe of these extraction wells. It is in these wells that we not only can try to collect some product but that there is also the potential to create hydraulic barriers (e.g., a BioWall™) using the natural bioplugging that is occurring. Once this barrier is in place, the permeability should drop dramatically and the plume is constrained; it cannot escape and go any further than it has. The redox front is a very positive aid to the generation of this situation. With more of an oxygen situation at work out there (e.g., by aeration), it is possible to "kick off" a better consortia of bacteria that will develop thick occlusive slimes which will store the hazardous product as it comes through. These material bioaccumulate and, if recalcitrant, can be removed using the BCHT™ to break up the slime plugs (and balls).

Another hazardous waste site that had severe biofouling problems was at an air force base in South Carolina. When the wells were camera-logged, there was 3 feet of JP4 jet fuel floating on the top of the aquifer. Recovery of the fuel by the extraction wells had gone down from 30 gallons per day per well to less than 3 (a 90% loss in recovery). I was requested to clean the thing up using BCHT™ because it would be cheaper than putting an airman out there with a bailer every day to collect the 3 gallons each day. When the camera went down into the well, there was 3 feet of JP4 jet fuel floating on the ground water. That fuel looked so pure, so pristine that it was almost like the fuel in their "auxiliary" storage tank. After treatment, we were taking pure fuel out of the ground and running it through two filters, putting the biocide back into it, and putting it back into the storage tanks. By the way, biocides are routinely used in jet fuel because biofouling (biocolloidal formation) is one of the leading causes of "flame-out" in an engine. The cause is bacterial growth as biofilms in the fuel tanks which then sloughs off (as biocolloids) and gets in the jet engine and plugs it! Afterward, it was rewarding to watch those A10 and F15 planes fly using the fuel that we had gotten back out of the ground.

Do not think that the JP4 or any other fuel always stays pure and pristine floating on an aquifer. When we looked at the diesel spill in Montana, using our interface probe and camera in the well, there was a heavy biomass equal in volume to the amount of fuel that we had observed floating on the ground water. Under the foot or so of fuel, there was a foot or so of biomass! Wells at the fringe of the plume were the first to plug up. You could call this "natural attenuation" at its best. The pure product well in the middle of the plume is usually the last one to go (biofoul), particularly if it is pumping strictly pure product. The fringe wells have always a little bit of a mix and some moisture has to be involved in it. At this site there were both aerobic and anaerobic forms of biofouling. The strictly anaerobic bacterial activity was dominated primarily by the sulfate reducing bacteria. It is a real pain to clean one of those fringe wells because, for a start, the bioplug is a hardened crusty mass that has built up *in situ* and is very hard to break up. If no preventative maintenance was done on a particular well, it was easy to lose 50% of the production (recovery). Those were found, by experience, to be difficult to bring back. The cheapest answer is just to drill another well, it is not worth the effort to try and rehabilitate those degrees of biofouling on a hazardous waste site. Again, the message is that preventative maintenance has to start as soon as the well is developed, whether it is used or not.

Remember that with monitoring wells, they are not a problem. It is the bacteria that are going to grow in, and around them, that are going to be the problem (primarily because of bioaccumulation, occlusion and biodegradation, while secondarily because of the false data that they then produce). Once, I had a call from the Savannah River plant where the nuclear trigger devices for H bombs, etc. were made and where a great deal of research has gone on. They had put in monitoring wells out in the swamp areas looking for "lost treasures". They were worried about quite a diverse range of solvents that had leaked out in that direction before the wells had been put in 10 years before. These wells were sampled once a year. The fellow from the Army's Corps of Engineers who was doing the sampling for them called me up and said, "I have got this black goop in my bailer when it came up! What is it?" The well had biofouled to such a point that he had eight 9-inch "nematodes" (or slime worms) growing down there. These wells were so badly biofouled that they were finally replaced with some new ones; here was an opportunity to yank up one of the old wells. It had a ball of black goo sliming all around the well and down into the sand aquifer that was above it. What had happened was that the water from the swamp had been leaking down the outside of the casings and going into the aquifer. Every sample they took out of that well over that 10-year sampling period was now suspect. In the water that had been sampled from these biofouled-monitoring wells, the traditional levels detected were in the parts per billion ranges, in the black goo around the monitoring well, those same hazardous materials were now measured to be in the percentile (%) scale. The black goo had been "bioagoomulating" these hazardous wastes! Ten years of data was in error due to the activities in the bioplugging around the monitoring well.

Things are often not what they seem and that appears to ring true in the hazardous waste industry. At another air force base, they thought they had collected all the jet fuel that had leaked out onto an aquifer. There was a clay structure in the aquifer that ran through a shelf or ridge underneath the runway. It was thought that this clay lens confined the aquifer. So sure were they of this that they did not bother to check on the integrity of that clay lens. Well, there was a crack in it. What was found was an area where the water running off the sand hills of South Carolina, heading down to the coastal plains never made it off to the swampy area. It was then decided to punch a hole out right where the water appeared to disappear. When the hole was drilled, another 300,000 gallons of aviation fuel was found floating on the aquifer. It had been running down into a fracture that was less than 4-inches wide. For over 30 years or how ever long they had been using jet fuel, it had been trickling down through that fracture in the lens to "pool" deeper down.

The monitoring well can biofoul. If there is any doubt at all about the data being gathered, take a sample and run a suite of BART™ tests on it; they are not that expensive. Most likely, the SRB, IRB, HAB and the SLYM will give you information as to whether the well is biofouling. Contact:

george@ARCC.net or roy.cullimore@uregina.ca

if you have difficulty understanding the BART™ data and interpreting it. You do not need a fancy analysis to determine whether there is biofouling, but if it is there, you should consider how relevant is it to the data gathered from the monitoring well?

Another case in point was March Air Force Base. They had a JP4 plume that went out under an adjacent town. I went to look at the wells at a time when they were in the process of replacing some of them. They were going to go with a maintenance program on the new well but 6 of the 9 extraction wells were being taken off-line. Three had been shut down and three more were on their way to shutting themselves down due to the excessive drawdowns in those wells. The project manager was concerned when he started to see vinyl chloride coming out of the wells. This was because there were very aggressive biofilms down there which were now going anaerobic. The vinyl chloride was being produced as a degradation product from trichlorethylene (TCE) by the bacteria under these anaerobic conditions. The message is: "bacteria can really mess you up during a hazardous waste treatment, do not ignore them and their activities or they will ignore you!"

IV. *TITANIC*, The Connection between Rusticles and Clogging
(by D. Roy Cullimore)

When you look at the plugs and the plugging that George has talked about, a lot of these are bioconcretions. Bioconcretions are a form of living concrete. What is being realized is that the plugging that you see in water wells are really forms of biological concretions. When you get these forms of growth, it is easy to see the reasons why a plug or plugs or biofilms that form around a well can be so tight and so difficult to deal with. There is also a lot of iron in these growths. The types of iron found are ferric oxides and hydroxides and are commonly referred to as geothytes. Iron can commonly make up 15 to 35% (dry weight) of the inside mass of that plug.

Another type of plug that you may not see in water wells, but you do see on mild steel on ships, is rusticles. Rusticles are currently growing on the *Titanic*. This ship sank in 1912 and has been the subject of much controversy along with historical and scientific interest. In 1996, I had the privilege of diving down to inspect the ship and the rusticles growing there. What is amazing about those growths is that they actually do bear the same structures and chemistry as the plugs that form around a water well.

Life is a learning experience. When you stop learning, then life becomes a boring roller coaster ride from night to day. On April Fool's Day, or April 1, 1997, the bacteria in the laboratory played an April fool's trick on me! They did this by making themselves into little slimy submarines. When 0.1 g of soil saturated with oil was put into an HAB-BART™ that had been pre-charged with distilled water, there was an UP reaction in two days. The bacteria producing methane gas followed this. This gas would be entrapped around the slimy particles (a.k.a., biocolloid and submarines) that would go up and then down again. It took about 8 seconds for the submarines to go up and an equal time to come down. Obviously, these biocolloids were able to precisely adjust their density. This rising and falling of the slimy submarines taught me that the bacteria could be pretty sophisticated.

The Titanic, for its time, was very sophisticated. The reason for the dive was that the Discovery Channel was making a series of documentaries on the ship going from yesterday, through to today and looking at the "tomorrow". I was there to deal with the tomorrow. What are the rusticles doing to the ship, how long will the steel last intact? Those were some of the questions and I am still learning to address at least a part of the answers. The documentary was released in 1997 as "Titanic, Anatomy of a Disaster".

R.M.S. Titanic was travelling at 22 knots when it hit an iceberg. There were small localized impact points (no great slash in the hull), which perforated the hull in five places. These impacts were, unfortunately, all in different water-tight compartments. It is thought that the steel plates were high in phosphorus and manganese and may have been vulnerable to a cold embrittlement, or that the wrought iron rivets failed. Because of this, the steel plates may have been more easily perforated or torn at the points of impact. The ship sank to the bottom quickly with a tragic loss of life. The hull now sits on the seafloor in three major parts: the bow at the front, a sheered part of the mid-section, and the stern which is twisted round with the propellers facing the forward bow section. The bow of the ship looks docked from the front approach. The stern is a tangled mass of twisted metal formed in the agony of the impact delivered by some force that today is causing heated debate. The rudder of the stern was driven in first during the collision with the seafloor; those huge propeller shafts were pushed up 50 feet so that the propellers are almost touching the underside of the stern.

When you come up to the Titanic, the seafloor is relatively sterile, looking like the Sahara Desert of the Seas. There was a lot of "sea snow" and I equate this to the "well snow" that can be seen when camera-logging a well. When I approached the ship itself in the submarine *Nautile*, the ship can be seen to be coated in these bioconcretions called rusticles.

When approaching the bow stem, there are still areas where the steel is relatively in tact and not coated with rusticles. Upon touring the bow, it became clear that where the steel had been damaged (e.g., torn or embrittled) by stresses imposed by the sinking and the impact with the seafloor, it was there that the rusticles were growing. In other words, where there were fractures and tearing, that was where the rusticles were. These gaps and tears within the ship steel created by the damage were where the rusticles first attached and started to grow.

Like the growths around water wells, the rusticles around the ship also moved in the direction from which the water was flowing. The way the bow section came down and landed, literally, had the water flowing passed the ship at approximately 1.8 miles per hour as if it were still under power. The water was moving from the front to the back of the bow section. The nutrients in the seawater were thus moving over the surface of the ship and the rusticles were all pointing in the direction from which the water was coming. In some cases, the rusticles actually appeared to be pointing backwards because of an undercurrent on the hull producing a reversed cyclic flow.

On the port side of the ship, just behind the anchor, it has been weakened so much that there is a big buckle in the hull that has been caved inwards. These buckled plates are totally covered in rusticles. It looks like there is severe corrosion on the inside of the ship, and the buckling may be partially a result of that corrosion.

On the deck itself, the handrails are buckled over because something must have hammered them down. Rusticles are growing profusely along those rails. Probably when the ship hit the bottom and the water following the ship (down wash) pressed the ship down further, it may have buckled the rails over. Rusticles are growing everywhere. There are thick growths on many of the vertical steel surfaces; they are hanging down from the steel girders inside the ship's hull. The wooden decks have rotted and only the caulking remains. The ship is now very fragile. In 1995 Cameron set two submarines on the deck, but I would not recommend doing that today because the ship is losing iron at a calculated rate of 0.1 of a ton a day (and there is probably a one order of magnitude error in that statement!). The steel is losing the iron to the rusticles that are, in turn, spraying it out as a red dust. The mass balance for the iron shows that it is moving to the rusticles that hold between 15 to 35% of the body weight as iron. The red dust is being produced (in the laboratory aquaria) at a rate of between 0.015 and 0.02% of the bulk biomass per day. This red dust, when released, contains between 8 and 20% iron. It can, therefore, be seen that the iron fabricated by humans into a steel ship is being returned by rusticles to the oceans at a rate that will see the Titanic gradually disappear as a recognizable structure and Nature will regain the iron the humans stole!

Where there is brass that is still intact and entire, but where there is steel, then the rusticles have been growing. Anywhere that the steel has been under tension, that is where the rusticles growths are the thickest. There is a lot of rotting going on. There are large holes and gaps in the bow of the ship. Inside, a general decay can be seen. Where there is bulging in the ship, the rusticles grow more extensively (some extend for 60 feet or more). You are wondering what the mass of the rusticles on the bow section would be. First calculations place the mass of rusticles that are visible at 650 tons (1996 estimate) and probably another 1,000 tons within the bow section. Every day those rusticles are growing bigger and throwing more of the iron out as a red dust.

Estimated Life Span of the Bow Section of *RMS Titanic* As a Recognizable Structure:

A. 60 years with iron losses of 0.1 ton/day
B. 400 years with iron losses of 1.0 ton/day
C. 6,000 years with iron losses of 0.01 ton/day

Figure Nine

The whole ship has become a hanging garden of rusticles. If you can imagine water wells that are not well maintained, they would be similar sites for the formation of hanging gardens of rusticles (we call that biofouling, plugging or bioplugging). The chemistry is very similar at the Titanic and down water wells except for the salt concentrations, but I have found that between 0.1 and 5.0% salt does not affect bacteria too much. They start to "stress out" at 6 to 8% salt.

The dominant organisms in bioplugging and rusticles are the iron-related bacteria with sulfate reducing and heterotrophic bacteria present. There are also more bacteria that can produce methane (biogas) and these can also grow down in water wells that are biofouling as well as in rusticles. They are all growing as different consortia at different locations in the rusticles.

What amazes me is the scale of the microbial "universes". There can be a whole variety of different consortia operating under very different environmental conditions all within a matter of millimeters!

We often forget the compressed scale of the environments within which these diverse consortia can function.

Dominant Bacterial Consortia Recovered from Rusticles:

- ☐ **Iron Related Bacteria**
- ☐ **Sulfate Reducing Bacteria**
- ☐ **Denitrifying Bacteria**
- ☐ **Methanogenic (Biogas producing) Bacteria**

They tend to cluster at different locations within the body of the rusticle.

Figure Ten

When rusticles were brought up to the surface, it had to be remembered that the pressure down here is 3.5 tons per square inch. The temperature is 1°C and the salt concentration is 4.2%. That would seem to be a very alien environment for us on the surface of this planet, but when we brought the rusticles up, they did survive. They did not appear to suffer from the "bends" and adapted well to the "extreme" environment (for them) present on the surface where we humans live.

From the Titanic experience and the more than quarter of a century spent working on biofouled water wells, all that I can say is that nature is a single integrated structure. The steel down a water well and the steel on the Titanic are both being infested by similar growths. They manifest themselves in different ways, but the microorganisms in the consortia are from the same genera, function in similar ways, and generate complex structures that reflect the environment. As Beijerinck stated more than a century ago: "Everything is everywhere, the environment selects". From all of the experiences that I have had, that is a most incisive and basic statement. Wells of any type are sited within Nature's integrated structures and, because of this, they are certainly going to become a focus for adaptation and activity. To manage such systems, there has to be unbiased observation, interpretation, and manipulation and then finally control. This then becomes a management strategy and that has to be sustainable.

APPENDIX

DEFINITION OF TERMS

These terms are given in alphabetic order to allow the reader to easily find the terms of interest. Note that the terms are also listed in the index.

Acidotrophic bacteria, bacteria which are able to flourish in very acidic (pH <3.5) conditions. Many are aerobic and function over relatively narrow pH ranges.
Adaptability, the ability of microorganisms either as individual strains or as a consortium of strains, to adapt to function in some way within a given environment. Often there is a lag (induction) time before this activity commences.
Aerobic microorganisms, microbes which can function using oxygen in their respiratory activities.
Aggressivity, the state in which an organism is active in its environment and able to compete with other strains for space, nutrients, water and gases.
Anaerobic microorganisms, microbes which are able to function in the absence of oxygen. Many of these organisms are able to function using oxygen when available (facultative anaerobes), while other strains cannot function in the presence of oxygen which is toxic to them (strictly anaerobic). There are a very few strict anaerobes which are not sensitive to the toxic effects of oxygen. These are known as aerotolerant.
Archaebacteria, are a group of bacteria which evolved very early on in the evolution of the planet. These bacteria are now found populating some of the extreme environments (e.g., highly saline, sulfur-rich, methane generating and high temperature).
Attachment, the act of a bacteria or a biocolloid becoming fixed to a surface. Growth may then follow leading to the formation of a biofilm.
Bacteriophage, a virus which infects bacteria and multiplies within the cells. Usually a bacteriophage can only infect a limited range of bacterial strains.
BART^{TM}, a patented biological activity reaction test biodetection system which can be customized to determine the aggressivity and composition of selected consortia of microorganisms.
BCHT^{TM}, a patented blended chemical heat treatment system which can be applied to rehabilitate plugged water wells and systems using a tri-phasic technology.
Bioaccumulator, a biological entity which is able to accumulate (either actively for degradation, or passively) chemicals within the surface coatings of EPS and/or within the cells themselves.

Bioamplifier, an organism which is able to catalyze a particular physical and/or chemical event, causing the event to occur at an accelerated rate.

Biocides, specific chemicals or compounds which have a deleterious impact (frequently lethal) on the targeted organism.

Biocolloid, a buoyant particle which is composed mostly of water bound together by EPS and populated by some microorganisms. Sizes may range from 6 to 100 microns or more in diameter. These suspended particles are also found to be able to act as bioaccumulators.

Biodegradation, the act of degrading a molecule to one or more smaller molecules by biochemical mechanisms (e.g., enzyme action).

Biodetector, an instrument, device or mechanism by which the presence of biological activity can be determined.

Bioencumbancy, the fraction of the volume within a biocolloid or biofilm occupied by viable cells.

Biofilm, an slime-like matrix composed of EPS within which one or more consortium of microorganisms flourish. These biofilms may either grow over surfaces, or occupy voids in a porous medium.

Biofouling, any deleterious event in which a definable biological activity causes a deterioration in an engineered or natural process or system. Deleterious effects range from plugging, corrosion, and plugging to gas production and bioaccumulation.

Biomass, the mass of a living entity which may be expressed as either the wet or dry weight. Biomass may, furthermore, be given as the total mass including all associated mass; or as the viable mass which would include just the viable cells. In biofilms, the total mass would relate to the total weight of the "slime" as such, while the viable mass would include just the mass directly associable with the living cells.

Biosensor, a device or methodology which utilizes the shift in a targeted signal (commonly electro-magnetic) to quantify a biological activity or presence.

Biozone, a localized site where a specific form of microbial consortium can be located.

Plugging, the generation of a mass which interferes with the physical functioning (e.g., hydraulic conductivity) of a porous medium (e.g., gravel pack, sand filter). Plugging can be formed through the maturation of biofilms fouling the media and may become complex in structure.

Plugging Risk Index (CRI), a factorial presentation of the likelihood of a significant plugging event occurring within a defined system.

Coliform bacteria, the presence of these bacteria is generally regarded as being indicative of an increased hygiene risk because of the potential for faecal contamination. The coliform bacteria are abundant in the faeces of warm blooded animals, and *Escherichia coli* is particularly common in the human species. Generally, the coliform bacteria do not survive long in natural waters and so form a good indicator organism for recent (significant) pollution due to raw or partially treated sewage.

Colony Forming Units (cfu), when microorganisms do grow on agar media they commonly form visible distinguishable structures composed mainly of cellular material which are called colonies. Each of these colonies is considered to have formed from a single colony-forming unit which may be a single cell or a clump of cells. By appropriate mathematical relationships of the dilution of the sample and the area of the agar inoculated, it is possible to predict a population as either cfu/ml (for liquids), cfu/g (for solids), or cfu/cm^2 (for surfaces).

Corrosion, the process of erosive deterioration in the physical form and engineered characteristics of a structure. These processes frequently involve electrolytic and/or corrosive chemical (e.g., acids) effects which are sometimes mediated by microbial activities. It has been observed that corrosive pitting can form directly under biofilms.

Culture, (verb) the act of successfully growing a unique strain or a consortium of microorganisms; (noun) a viable collection of a single strain of microorganisms which has been selectively grown.

Denitrification, the process of reducing nitrate via nitrite to nitrogen gas by bacterial action. There are four stages in this process. In water which has become polluted with sources of organic nitrogen (e.g., sewage or septic waste) and has been subsequently subjected to aerobic (oxidative) nitrification, nitrates are a major product. If conditions now become anaerobic, these accumulating nitrates are reduced by denitrification.

Disinfection, the act of destroying by chemical and/or physical means microorganisms that are causing an undesirable infestation or effect at a given site. It does not mean that all microorganisms are killed, it means that there is a selective action.

Encrustation, a relatively solid plate-like or crystalline structure coating a surface. It appears to be chemical in nature due to the hardness of the structure. Often brittle (when dry) or plastic (when wet), the organic content is usually relatively small.

EPS or extracellular polymeric substances, many microorganisms do produce an "overcoat" of polymers outside of the cell. These polymers bind water and various chemicals to form protective and storage functions.

Eutrophic conditions, these occur when there is an abundance of nutrients and the microorganisms are able to grow to form a large biomass. A rapidly plugging well due to IRB growth could be considered as being a eutrophic event.

Fringe Effects, the zone wherein the treatment is marginalized and, therefore, has a lesser or different impact.

Gallionella, is a well-known iron-related bacterium which is easily recognized by the long often twisted ribbon-like tail they produce. These tails will break off and are carried with the water flow.

Gram Stain (gRAM), is a standard staining procedure which is frequently used as one of the first stages in the identification of bacteria into Gram negative (pink-red) and Gram positive (blue-purple) types.

Halotrophs, microorganisms able to survive and grow in brine solutions. Some of these organisms cannot even survive when salt concentration are less than 12%.

Heterotrophic microorganisms, those microbes which obtain their energy from the breaking down of organic material. Some of these microbes are very specialized (e.g., cellulose degraders) while other can utilize a variety of organic compounds.

Hydrolysis, the act degrading complex molecules (e.g., polymers) into smaller molecules through reactions involving H^+ and OH^-.

Incubation, the act of growing an organism under conditions that will encourage rapid growth (compared to natural conditions).

Induction period, the period of adaptation that an organism has to pass through before it is able to flourish in a favorable habitat.

Infiltration, the act of a material or organism passaging into a porous medium.

In situ, at site.

Invasiveness, the ability of an organism to enter into an environment and function at some level between survival and rapid growth.

In vitro, under controlled (laboratory) conditions.

Iron-oxidizing bacteria, those bacteria able to oxidize iron by any means from a reduced form of iron (Iron (II), Fe^{++}, Ferrous form) to an oxidized (Iron (III), Fe^{+++}, Ferric) state.

Iron-reducing bacteria, those bacteria which are able to reduce iron by any means from an oxidized form (ferric) to a reduced (ferrous) state.

Iron-related bacteria (IRB), all of those bacteria which are able to accumulate iron in any form beyond that required for basic metabolic functioning. These accumulated iron compounds generally collect within the slime (EPS) around the cells and gradually harden (shift from amorphous to crystalline forms) over time.

Limiting nutrient, a major nutrient which is in short supply and, thereby, restricts the growth of a biomass. Limitations could also be created by the limiting nutrient distorting the ratios of nutritional elements to beyond that range which would support growth.

Macrofouling, an intense and/or widespread form of biofouling.

Magnetotactic bacteria, a group of bacteria which actually possess "biocompasses" (called magnetosomes). These bacteria are able to orient themselves within electromagnetic fields.

Marginal plugging, where there is less than a 20% loss in production capacity but a water well shows some symptoms of being plugged. This phenomenon is considered marginal but discernable.

Mechanical disruption, the use of physical methods (such as freezing, sonication, pressure pulses, radical thermal gradients) to disrupt a biofouling event.

Membrane filtration (MF), the use of a non-absorbent porous membrane to trap particles (including bacteria) which allow the water to filter through. It is a technique used to enumerate low numbers of bacteria in water by concentrating the cells onto the filter's surface where they may be grown to form visible countable colonies. Pore sizes commonly employed are of 0.22 and 0.45 microns diameter.

Mesotroph, an organism which will grow over a temperature range somewhere between 15 and 45°C.

Microbial growth potential (MGP), the theoretical growth which may be expected to occur within a defined environment utilizing the available nutrients at a maximum efficiency.

Microcosm, a habitat within which there is little diversity in the organisms present or the environmental factors. Often applied to laboratory simulations of "real world" situations (e.g., a well microcosm used to simulate plugging processes in wells).

MPN, most probable number, to determine populations, some microbiological techniques use a statistical projection of the population as the most probable number and do not specifically count the individual cells as such.

Negative staining, sometimes bacteria are not easy to stain due to the fuzzy EPS around the cell. One alternative is to stain the background so that the cell (and its EPS slimes) can be more easily viewed microscopically.

Nitrification, biological conversion of ammonium to nitrate occurs under oxidative (aerobic) conditions. It is a major part of the nitrogen cycle.

Nitrogen fixation, the act of a biological system fixing nitrogen usually to ammonium (as the intermediary). Some microorganisms under stress imposed by an inadequate nitrogen resource for growth can fixate nitrogen.

Nosocomial, term used for microorganisms which are normal inhabitants of a natural environment, but can, under certain circumstances, cause infections in warm blooded animals including man.

Occlusion, the reduction of hydraulic conductivity (flow) through a porous medium as a result of the growth of a plugging structure which is now occupying a significant part of the void volume.

Oligotrophic conditions, these occur where there are few nutrients in the system and the microorganisms are not able to grow to form a significant biomass.

Particulates, suspended material in water, may be inorganic and/or organic in nature. May contain living microorganisms and be colloidal in structure (biocolloid).

Pasteurization, the process of applying heat to a sufficient extent to retard or destroy a recognized nuisance microbial population. Usually involves the application of heat to a minimum rise of 40 C° above the ambient temperature for the system.

Planktonic, references microorganisms able to grow while suspended independently in water.

Plugging, the act of particulate and/or biofilms filling void spaces within a porous medium to significantly reduce water flow through that medium.

Preventative Maintenance (PM), a management strategy to allow the ongoing monitoring of a system or process to ensure that there is a reactive scenario in place to control any recurrence deterioration.

Pseudomonad bacteria, are Gram negative aerobic bacteria which frequently dominate in waters polluted with specific organics. Some strains are nosocomial pathogens.

Recalcitrant, a chemical which does not degrade, is stable, and persists within the environment.

Rehabilitation, the returning of a well or other system to its original specified state by the application of suitable treatments.

Reinfection, the act of an infestation becoming reestablished within a system or process.

Sessile, organisms that are attached to a surface either directly or indirectly.

Shock treatment, the application of a higher than normal dose in order to maximize the effectiveness of the treatment being applied.

Slime, a surface growth on, or originating from, a surface. It may be jelly-like in form. Such slimes are usually infested with various microorganisms and can act as sites for the bioaccumulation of various chemicals.

Slime Forming Bacteria (SLYM), bacteria which do produce slimes (from EPS) but do not necessarily accumulate iron within these slimes.

Sloughing, the act of a slime, for whatever reasons, breaking up and releasing particles (from the slime) into the water passing over the slime.

Spreadplate, name given to the microbiological procedure for enumerating microorganisms through their ability to form colonies on selected agar media when dispersed ("spread") over the agar surface and incubated.

Sulfate-Reducing Bacteria (SRB), anaerobic bacteria which are able to reduce sulfate to hydrogen sulfide. This event may initiate electrolytic corrosion and/or rotten egg taste and odors in water.

Thermal Death Point, the lowest temperature that is required to destroy a specific strain or consortium of microorganisms in ten minutes.

Thermotroph, an organism which is able to grow at temperatures in excess of $45°C$.

Total Nitrogen, the total amount of nitrogenous compounds determined to be in the water exclusive of nitrogen (N_2) itself. Major fractions include nitrate - N, nitrite - N, ammonium - N and Kjeldahl nitrogen.

Total Organic Carbon, the total amount of organic carbon in the sample. May include soluble and particulate forms which may, or may not be, recalcitrant.

Total Phosphorus, the total amount of phosphorus detected in the sample. May be four forms: soluble inorganic phosphorus, SIP; soluble organic phosphorus, SOP; particulate inorganic phosphorus, PIP; and particulate organic phosphorus, POP.

Tubercles, these are raised encrustations often incorporating rusty flakes. They commonly grow on surfaces and form the sites for enhanced electrolytic corrosion. Biofilms are often generated within the tubercle.

Tyndallization, the act of repeating a treatment sequentially in order to destroy the survivors of the previous treatment as they grow and become more vulnerable to the treatment process. Commonly, the treatment is performed three times with a sufficient interval between to allow the survivors to grow.

Ultramicrobacteria (UMB), very small, electrically neutral microorganisms which are in a state of suspended animation. They are able to survive in this form for very long periods of time and recover when exposed to a favorable environment.

Viable units, a form of measuring the number of microorganisms in terms of their viable (detectable) units which may consist of one or more cells.

Wolfe's medium, a specialized medium widely used for the selective growth of *Gallionella*.

WR medium, a modified Winogradsky's medium used to determine the presence and numbers of iron-related bacteria.

SELECTED BIBLIOGRAPHY

Anonymous, *Problem Organisms in Water: Identification and Treatment,* American Water Works Association, Denver, 1995
Baron, E.J., Chang, R.S., Howard, D.H., Miller, J.N. and J.A. Turner, *Medical Microbiology, A Short Course,* Wiley-Liss, New York, 1994
Bouwer, H., *Groundwater Hydrology,* McGraw-Hill, Inc., New York, 1978
Buchanan, B., De La Cruz, Macpherson, J., and K. Williamson, *Water Wells ... that last for generations,* Alberta Agriculture, Food and Rural development, Edmonton, 1996
Chapelle, F.H., *Ground-Water Microbiology and Geochemistry,* John Wiley & Sons, New York, 1992
Cullimore, D.R. and A.E. McCann, The Identification, Cultivation, and Control of Iron Bacteria in Ground Water. In *Aquatic Microbiology.* Skinner, F.A. and J.M. Shewan, Eds, Academic Press Inc., New York, 1977
Cullimore, D.R., Ed. *IPSCO 1986 Think Tank on Biofilms and Biofouling in Wells and Groundwater Systems,* Regina Water Research Institute, University of Regina, 1987
Cullimore, D.R., *Practical Manual of Bacterial Identification,* Regina Water Research Institute, University of Regina, 1996
Cullimore, D.R., *Practical Manual of Groundwater Microbiology,* Lewis Publishers, Boca Raton, 1993
Driscoll, F.G., Ed., *Groundwater and Wells, second edition,* Johnson Division, St. Paul, Minnesota, 1986
Ellis, D., *Iron Bacteria,* Methuen & Co. Ltd., London, 1919.
Hattori, T., *The Viable Count Quantitative and Environmental Aspects,* Springer-Verlag, Berlin, 1988
Howsam, P., Ed., *Microbiology in Civil Engineering Federation of European Microbiology Societies Symposium No. 59,* E. & F.N.Spon, London, 1990
Howsam, P., Ed., *Water Wells Monitoring, Maintenance, Rehabilitation,* E. & F.N.Spon, London, 1990
Kissane, J.A. and R.E. Leach, *Redevelopment of Relief Wells, Upper Wood River Drainage and Levee District, Madison County, Illinois,* Technical Report REMR-GT-16, US Army's Corps of Engineers, Washington D.C., 1993
Kissane, J.A., *Use of New Well Redevelopment Techniques on Relief Wells in the Upper Wood River Drainage and Levee System,* The REMR Bulletin, 5(3), 1988
Krumbein, W.E., Ed., *Microbial Geochemistry,* Blackwell Scientific Publications, Oxford, 1983
Leach, R.E. and G. Hackett, *Evaluation of the Rehabilitation Program for Relief Wells at Leesville, Ohio,* Technical Report REMR-GT-18, US Army's Corps of Engineers, 1992
Ledin, M. and K. Pedersen, *The Environmental Impact of Mine Waste – Role of Microorganisms and their Significance in Treatment of Mine Wastes, Report AFR-Report 83,* Swedish Waste Research Council, Stockholm, 1995
Mansuy, N., *Water Well Rehabilitation, A Practical Guide to Understanding Well Problems and Solutions,* Lewis Publishers, Boca Raton, 1998
Marston, D.L., *Law for Professional Engineers, second edition,* McGraw-Hill Ryerson Limited, Toronto, 1985

Nielsen, D.M., Ed., *Practical Handbook of Ground-water Monitoring,* Lewis Publishers, Boca Raton, 1991

Pipes, W.O., Ed., *Bacterial Indicators of Pollution,* CRC Press, Boca Raton 1982

Postgate, J., *Microbes and Man, Third Edition,* Cambridge University Press, Cambridge, 1992.

Rodina, A.G., *Methods in Aquatic Microbiology,* University Park Press, Baltimore, 1972

Rogers, B. and R. Leach, *Use of Blended Chemical Heat Treatment (BCHT) Procedure to Clean Contaminated Wells on Superfund Sites,* The REMR Bulletin, 10(3), 1993

Sawyer, C.N., McCarty, P.L. and G.P, Parkin *Chemistry for Civil Engineering, Fourth edition,* McGraw-Hill Book Company, New York, 1994

Smith, S. A., *Monitoring and Remediation Wells Problem Prevention, Maintenance and Rehabilitation,* Lewis Publishers, Boca Raton, 1995

Smith, S.A., *Methods for Monitoring Iron and Manganese Biofouling in Water Supply Wells,* AWWA Research Foundation, Denver, 1992

Wells, S., *Titanic, Legacy of the World's Greatest Ocean Liner,* Tahabi Books, 1997

Wolfe, R.S., Cultivation, Morphology, and Classification of Iron Bacteria, *J. AWWA,* 42:849-858, 1958

Index

A

ammonium · 18, 31
applications protocol · 113
Armstrong scenario · 46

B

backwashing · 69
BART-SOFT™ · 131
BART™ · 50, 71, 88, 93, 152
basic rules · 122
BCHT™ · 21, 58, 59, 60, 61, 62, 64, 72, 73, 74, 75, 85, 96, 97, 101, 103, 104, 105, 106, 107, 108, 109, 110, 111, 112, 113, 114, 115, 116, 118, 119, 120, 121, 125, 126, 130, 132, 133, 139, 143, 153, 156, 158, 159
Beijerinck · 166
Bioaccumulation · 84, 155
biocolloids · 13, 16, 20, 33, 67, 93, 159, 162
biofouling process · 37
biological challenge · 10, 19
bioplume · 56, 154, 156
Bofers Site · 83
Boiler · 106
boils · 86, 109, 138
Brookville Lake · 64

C

carbonates · 29, 58
causes of biofouling · 22, 34
CB-4 · 59, 105, 106, 130, 139
clogging · 54, 162
cloudiness · 4, 5
coliform · 2, 6, 9, 18, 19, 34, 53, 55, 120, 121, 129
coliform problem · 129
colonization · 37
color · 4, 5
cone of influence · 137, 138, 145
covert plugging · 48
crystalline chemical complexes · 25

D

diagnosis of biofouling · 119
disinfectant · 21, 59, 63, 68, 78, 90, 101, 110, 129, 145
dispersants · 10, 21
disruption · 60, 61, 64, 66, 94, 104, 110, 111, 112
double gravel pack · 98
drawdown · 3, 45

E

earthy-musty odors · 6, 49
encrustations · 3, 4, 5, 9, 20, 26, 51, 58, 65, 78, 84, 140, 143, 157
environmental concerns · 8
EPS · 20, 25, 26

F

fishy odors · 5
flip-flop · 97, 130, 151
freezing · 16
fruity odors · 7

G

Gallionella ferruginea · 52
Garrison Dam · 70
geothite · 29
geyser effects · 104

H

hardened plates · 48
hazard waste sites · 116, 126
hematite · 29
historical background · 128
HTH · 63, 64, 88, 89, 90, 91, 94, 142
hydraulic barriers · 159
hydrology · 9, 151
hydroxyacetic acid · 78, 158
hygiene problems · 3

I

iron · 11, 29, 30, 40, 51, 81, 162, 166
iron-related bacteria · 4, 51, 52, 55, 63, 67, 77, 165

J

jetting · 59, 66, 102, 103, 104, 106, 107, 108, 109, 112, 115, 118, 132, 133, 135, 136
JP-4 · 66, 67, 69

K

kerosene-like odors · 6

L

Leesville · 75
Leptothrix-Sphearotilus group · 77
linear polyphosphate · 78, 81

M

major symptoms · 23
management of biofouling · 2
manganese · 12, 29, 40
mean particle size · 42, 43
mesocosms · 31, 55, 58
Mississippi River · 85
mucoid tubercles · 47

N

natural attenuation · 127
nitrate · 18, 31, 32
nitrite · 18, 32
nozzles · 107, 112, 115
nutrient loadings · 19
nutrients · 17, 50
NuWell® · 149

O

odor · 3, 5, 6, 7, 14, 53, 120
organic contaminant plume · 4

P

pasteurization · 54
PFRA-TS · 53, 97, 119, 150, 152, 155
pH · 13, 14, 15, 21, 58, 60, 64, 68, 70, 73, 86, 90, 94, 97, 100, 102, 124, 126, 130, 142, 149, 151, 154
phosphate · 18, 32, 37, 63, 78, 79, 83, 126
physical collapse · 9
plug formation · 24, 27
plugging · 1, 3, 4, 5, 9, 11, 14, 24, 26, 27, 28, 33, 34, 35, 37, 39, 40, 41, 42, 44, 45, 46, 47, 48, 49, 50, 51, 55, 56, 59, 61, 62, 64, 65, 66, 67, 68, 69, 71, 75, 77, 84, 85, 86, 96, 97, 98, 101, 117, 121, 127, 139, 141, 143, 145, 154, 155, 157, 162, 165
pollutants · 9
polyphosphates · 18, 33, 79, 83, 128
preventative maintenance · 2, 7, 9, 11, 12, 22, 23, 24, 35, 37, 43, 64, 69, 71, 83, 85, 101, 105, 119, 120, 124, 125, 129, 130, 132, 146, 147, 158, 160
proteins · 18
psychrotrophs · 16, 45
pump rates · 45

R

recalcitrant · 28
record keeping · 130
recovery wells · 66, 67, 69
redevelopment · 91, 92, 93
redox · 14, 33, 46, 81
redox front · 14, 17, 27, 28, 29, 32, 33, 34, 36, 37, 46, 47, 48, 159
rehabilitation · 9, 22, 24, 28, 35, 36, 39, 43, 51, 54, 61, 64, 65, 66, 69, 74, 75, 79, 83, 85, 91, 99, 102, 103, 113, 114, 118, 119, 120, 128, 130, 132, 133, 141, 145, 147, 150, 151, 152, 153, 154
rehabilitation techniques · 8
risk-assessment · 9
rotten egg odor · 5
rusticles · 162, 164, 166

S

sand boils · 85
satellite well · 125, 126
sea snow · 163

septic odors · 6
septic wastewater · 2, 4
sequence · 27
Shaw Air Force Base · 66
skunky · 7
slime · 5, 12, 13, 14, 20, 24, 25, 26, 38, 55, 71, 72, 74, 86, 117, 158, 159, 160
slime formations · 9, 68
specific capacity · 80, 92, 93, 94
spontaneous recoveries · 28
spot treatment · 136
stages of biofouling · 39
subsurface microbiology · 9
successful rehabilitation · 109
sulfamic acid · 64, 68, 72, 89, 102, 105, 110, 141, 142, 149
sulfate-reducing bacteria · 5
surge blocks · 98, 139
sustainable water well initiative · 134, 155

T

taste · 3, 5, 14, 53
temperature · 15, 34, 44
thermotrophic · 16, 45
Thiobacillus · See
THM · 101, 110
Titanic · 162
total phosphorus · 32, 40
total suspended solids · 12, 42

U

Upper Woods River · 62

V

vegetable odors · 6
video-logging · 65, 71, 72, 74
void volume occupancy · 38

W

water production · 10, 11, 36
Waverly · 120, 123, 156
well snow · 163